Transforming Organizations

One Process at a Time

T0303851

Continuous Improvement Series

Series Editors:
Elizabeth A. Cudney and Tina Kanti Agustiady

PUBLISHED TITLES

Affordability: Integrating Value, Customer, and Cost for Continuous Improvement
Paul Walter Odomirok, Sr.

Continuous Improvement, Probability, and Statistics: Using Creative Hands-On Techniques
William Hooper

Design for Six Sigma: A Practical Approach through Innovation
Elizabeth A. Cudney and Tina Kanti Agustiady

FORTHCOMING TITLES

Transforming Organizations: One Process at a Time
Kathryn A. LeRoy

Statistical Process Control: A Pragmatic Approach
Stephen Mundwiller

Robust Quality: Powerful Integration of Data Science and Process Engineering
Rajesh Jugulum

Building a Sustainable Lean Culture: An Implementation Guide
Tina Agustiady and Elizabeth A. Cudney

Transforming Organizations
One Process at a Time

Kathryn A. LeRoy

CRC Press
Taylor & Francis Group
Boca Raton London New York

CRC Press is an imprint of the
Taylor & Francis Group, an **informa** business

CRC Press
Taylor & Francis Group
6000 Broken Sound Parkway NW, Suite 300
Boca Raton, FL 33487-2742

© 2018 by Taylor & Francis Group, LLC
CRC Press is an imprint of Taylor & Francis Group, an Informa business

No claim to original U.S. Government works

International Standard Book Number-13: 978-1-138-19772-5 (Paperback)
978-1-138-29736-4 (Hardback)

Library of Congress Cataloging-in-Publication Data

Names: LeRoy, Kathryn A., author.
Title: Transforming organizations : one process at a time / Kathryn A. LeRoy.
Description: Boca Raton, FL : CRC Press, 2017. | Series: Continuous improvement series
Identifiers: LCCN 2017018674| ISBN 9781138197725 (pbk. : alk. paper) | ISBN 9781138297364 (hardback : alk. paper) | ISBN 9781315277165 (ebook)
Subjects: LCSH: Total quality management. | Continuous improvement process. | Organizational change.
Classification: LCC HD62.15 .L47 2017 | DDC 658.4/013--dc23
LC record available at https://lccn.loc.gov/2017018674

Visit the Taylor & Francis Web site at
http://www.taylorandfrancis.com

and the CRC Press Web site at
http://www.crcpress.com

Printed and bound in Great Britain by
TJ International Ltd, Padstow, Cornwall

To Gene, who believes in possibilities.

His encouragement, support, and love help me be my best every day.

Contents

Section three: Sustain transformation

Preface

What if? Those two simple words hang in the air and follow me. When it began may not matter as much as the fact that those two words never leave my side. I suspect that my colleagues, friends, and family have often wished they could pop the bubble—the one you see in cartoons that follows the character everywhere she goes.

Perhaps it started with my mother's words that still echo in my ears and my heart. "Do your best." So, what is my best? How will I ever know that I did "my best?" Everything we do presents an opportunity to improve. Maybe not today, but the world does not stand still. We can either adapt, change, or find ourselves forever frozen in a world that no longer exists.

Businesses across all sectors and of all sizes face this threat. Leaders must recognize the signs of atrophy. They must remain alert to the evolving landscape of a global society. We all have a responsibility to guide people through adapting to the realities of today. Our survival has and always will depend on whether we choose to see with eyes wide open and ask those two simple words, *what if?*

What if we asked our customers what they expect in a quality service or product? What if we had a strategy to grow our company and then carried out that plan? What if we created work environments with open and robust dialogue for decision-making? What if we knew if how our work met the requirements for a quality product for those we serve? What if we had a method to improve what does not work? What if we could transform every workplace so that everyone, every day, everywhere, could do their best?

There is no shortage of books on leadership, customer service, employee engagement, or process improvement. My bookshelves overflow with some of the best. This book emerged from my experiences in working with and within organizations who struggled with the *how*. How do we turn this ship around? How and where do we begin?

Transformation and excellence should not be difficult or out of reach. Jargon, complex models, and our own fear of not knowing what lies ahead, can overwhelm the most ardent change leaders. My goal focused

on seeking a balance between the background knowledge of key concepts and practical examples from life and from other professionals who
ask *what if* every day. These leaders, who shared their stories throughout this book, create cultures of excellence. They face similar challenges
and obstacles, but they relentlessly pursue excellence and help everyone
around them be and become their best.

Acknowledgments

Dr. Bryan Cole continues to be my teacher and colleague extraordinaire. Dr. Stan Hall always tells me the truth and remains a trusted thought partner. I have become a better teacher, facilitator, and person because of the many colleagues who share my passion for excellence. A few of them shared with you through *Insights for the Journey*, but there are many others who have crossed my path and contributed to my professional learning.

Families must put up with a writer who "gets in the zone" and forgets about dinner or makes children wait to play. Mine is no different. These are the special ones who make me smile and proud because they do their best every day: Gene, Gregory, Andrew, Juliane, Reid, Emily, and Dexter. Thanks to each of you!

Insights for the Journey
Interviewee List

The insights and expertise of each of these individuals contributed to making this book a practical guide for transforming workplaces. They represent education, health care, nonprofit, and business organizations and have one common goal: creating cultures of excellence. I learned from each of them and hope you find support, guidance, and inspiration from their experiences and practices.

Dr. Trent Beach is a recognized healthcare leader involved in transformational change and innovative patient-centered quality and value improvement in medication use systems. He actively pursues healthcare leadership, advocacy, systems improvement through organizations such as the Baldrige Performance Excellence Program, American College of Healthcare Executives, and American Society of Health-Systems Pharmacists.

Dr. Bryan R. Cole is Professor Emeritus of Educational Administration in the Department of Educational Administration and Human Resource Development at Texas A&M University. His professional interests include performance excellence in educational and non-profit systems, and he is a frequent speaker and consultant on organizational analysis, planning and managing complex organizational systems and implementation of continuous improvement.

Ben Copeland is the Assistant Superintendent of Operations and Administration for Lynchburg City Schools in Virginia and a retired U.S. Marine. He chose public education as his second career and is using his operational knowledge and educational experiences to enhance the lives of students and help organizations reach their full potential.

Dr. Shelby Danks is a thought leader in performance excellence and organizational assessment, and is currently a Managing Researcher at

McREL International. She has empowered more than 100 organizations through service on the Board of Examiners for the National Baldrige Performance Excellence Program, Panel of Judges for the Texas Award for Performance Excellence, the Peer Review Corp for the Higher Learning Commission, and as an accreditor for AdvancED. Her research interests include leadership behaviors and transparency, evidence-based decision making, and innovation management.

Genie Wilson Dillon is a quality improvement advisor, manager, and director in manufacturing, healthcare, education, and nonprofit sectors for more than 45 years. She is recognized as a Quality Texas Fellow and Baldrige Senior Examiner and key contributor in leadership roles for the Quality Texas Foundation and Alaska Performance Excellence Foundation.

Vic Figurelli is the founder of the Williamson County Institute for Performance Excellence in nonprofits.

Brian Francis is the executive director of the Texas Department of Licensing and Regulation (TDLR). TDLR oversees more than 712,000 licensees in 32 diverse occupations and industries. Brian leads this agency with a vision of excellence and a focus on asking the difficult questions necessary during a season of dynamic change in state government. Brian's high energy combined with an MBA and a poet's view of the world means he models the leadership needed to bring about the most positive future of regulation: efficient, cost-effective, and innovative.

Susan Fumo is the executive director of School Improvement in Rockford Public Schools, a district of 28,000 students. With 20 years of experience in public education, she oversees data-driven goal setting and action planning in 46 schools in a large, urban district.

Dr. Jeff Goldhorn is the executive director of Education Service Center, Region 20 (ESC-20) in San Antonio, Texas where he provides leadership in pursuit of the organization's mission of impacting education through high quality, cost effective products and services. Jeff attributes much of the success of ESC-20 to the culture of excellence that has been established through a mindset of continuous improvement.

Zachary Haines is vice president, Information Technology and Performance Excellence for Goodwill Central Texas. Haines led the founding of the Performance Excellence Department and spearheaded adoption of the Baldrige criteria as a model for organizational improvement, which helped the organization achieve the 2013 Texas Award for Performance Excellence.

Frankie Jackson is the associate superintendent and chief technology officer at Cypress Fairbanks ISD, outside of Houston, Texas serving over 115,000 students, the 22nd largest school district in the nation. With 23 years of public education experience, her expertise lies in Technology Leadership, Communication, and Transformation.

Debra L. Kosarek has more than 30 years of experience in both the private and public sectors. She has helped organizations continuously improve through planning, systems improvement, and developing human resources.

Laura Longmire is founder and partner at Strategic Quality Initiatives. She is an executive leader in the fields of Quality and Human Resources with key expertise in process management, benchmarking, and knowledge management. Laura has served as an examiner, trainer, quality fellow and judge for the Texas Award for Performance Excellence. In the field of quality, she has been a leader of an award recipient for the Malcolm Baldrige National Quality Award, a Baldrige examiner, AME assessor, and the panel of judges for the Gryna Award.

Dr. Kelly Monson currently serves at the Chief Continuous Improvement Officer for the Rockford Public Schools (RPS205) in northern Illinois. With over 20 years of experience in education, she oversees strategic planning, district-wide processes, systems training, and data analysis for a large urban district.

Dr. C. Ryan Oakley founded Spring Creek Dentistry and provides comprehensive dental service for families in the Houston area. Dr. Oakley, a believer in lifelong education, has completed several hundred hours of continuing education and has earned fellowship status with the Academy of General Dentistry. He genuinely cares about his patients and enjoys learning about their lives, dreams, and health.

Michél Patterson is a senior executive and transformation leader, holds a master of science in statistics, is certified as a Lean Six Sigma Master Black Belt, and advanced practitioner in applied behavior science. She has more than 30 years of operational excellence and continuous improvement experience in a variety of industries including manufacturing, technology, service, and government.

Anna Prow designs and directs integrated change campaigns for nonprofits of all sizes in term engagements of 12–36 months. She's both expert and practitioner with unparalleled experience from diverse sectors and issue areas, including nearly 15 years as a social sector executive.

Rick Rozelle is president and chief information officer (CIO) for the Center for Educational Leadership and Technology (CELT). Founded in 1991, CELT is a nationally recognized information technology architect and business/learning systems integrator for K–12 education. He specializes in helping districts with strategic planning, strategic project oversight, process improvement/reengineering, enterprise architecture, and data governance. He has also served as CIO in learning organizations including Charlotte-Mecklenburg Schools and the Tennessee Department of Education.

Dr. Cynthia St. John is an executive consultant and coach for C-level leaders, and is the former chief learning officer for one of the largest faith-based healthcare systems in the country. Dr. St. John has led large scale corporate transformations, understands the factors most critical to success, and now applies this knowledge to help simplify success for other senior leaders and their organizations.

Dr. JoAnn Sternke is the retired superintendent of schools for the Pewaukee School District in Wisconsin. The Pewaukee School District has been dedicated to using the Baldrige Criteria for Performance Excellence and, in 2013, was honored to receive the Malcolm Baldrige National Quality Award. With this commitment, the district pursues a relentless focus on using strategic planning, results data, and key work process identification to leverage improvement.

Doug Waldorf has worked in business and health care for more than 15 years. He serves on the Board of Trustees for Liberty Tech Charter School as well as on the Board of Examiners for the Malcolm Baldrige National Quality Award.

Author

Kathryn A. LeRoy, PhD, CMQ/OE, has a lifelong passion for excellence.

Her goal is to help leaders at all levels create environments where everyone, every day, can be and become their best.

Kathryn's professional journey has taken her down varied paths. She has traveled from classroom teacher to champion for the transformation of schools and workplaces. In every role, one constant shapes her work—remaining a learner.

Kathryn believes that we can change organizations one process at a time if we are willing to learn, and if we choose to believe in possibilities. She invites—and challenges—you to join her to make it happen.

Kathryn is an American Society for Quality Certified Quality Manager for Organizational Excellence. She has served as a senior member of the Board of Examiners for the Malcolm Baldrige National Quality Award and the Texas Award for Performance Excellence. Kathryn also volunteered as vice-chair of the Board of Overseers, team leader, feedback writer, process coach, and trainer for the Quality Texas Foundation. She earned a BA in English from the University of Houston, an MEd in curriculum and instruction from the University of Houston, and a PhD in educational administration from Texas A&M University.

chapter one

Transform through a culture of excellence

> Knowing is not enough; we must apply.
> Willing is not enough; we must do.[1]

Goethe

Do you want to transform your organization and create a culture of excellence? High performing teams and organizations possess the desire and persistence to achieve at the highest level—essential to creating a culture of excellence. Are we willing to learn, to persist, to be courageous? The first lesson in this book is that transformation begins with *doing*.

Act to transform

I have observed organizations that try to think as a system. They try to continually improve. They try to engage employees. They try to measure effectiveness of their work. They try to listen and understand the needs of their customers and stakeholders. They try; trying is not doing.

Doing and *acting* build a culture of excellence. Creating that culture requires keeping "a constant awareness and vigilance to always be your best with passion, competency, flexibility, communication, and ownership."[2] Passion is not about platitudes or posters exhorting employees to strive for excellence. Passion is about inspiring a positive focus on possibilities. Too often, we become our own worst enemy simply because we lose hope and become fixated on what appear as insurmountable odds.

Transformation of any kind requires many types of skills across the organization. Building the capacity of everyone to accomplish the work is essential for excellence. We must not limit competency to professional skills. Interpersonal skills and emotional intelligence support effective decision making, collaboration, and become the foundation for a culture of excellence. As we become more confident in our ability to do our best, individually and in teams, we can tackle the seemingly impossible.

Communicate, communicate, communicate! Communication in a culture of excellence balances listening and speaking. Whether you have a large organization or your business consists of you and one or two others, never underestimate the importance of effective communication. The way we deliver the message and the extent to which we listen and respond to others has the power to shape perceptions. A respected colleague pointed out to me the interdependent relationship between communication, commitment, and compromise. Success often depends on our willingness to communicate our commitment knowing that compromise influences our relationships and decisions. This one area can derail our efforts, and we can easily delude ourselves in thinking that *everyone knows* and *everyone shares the same commitment.*

Without exception, continuous improvement requires change. When you begin to transform your organization and improve how you do your work, the status quo does not exist. Inherent to excellence is the continual recalibration of our knowledge and skills and the way we achieve goals or outcomes. The difference between success and failure often lies in the extent to which you and your team can persist, flex with the inevitable, and accept that doing your best always results in change.

Blame undermines relationships and progress, leaves us frustrated, and erodes trust. We cannot blame others for our circumstances. The complexity of modern organizations tempts us to approach our situation as victims. Operating as a victim creates blind spots where we fail to recognize the central issue, or that we likely have created the crisis and the only way out is to own it. We may balk at the word accountability as overused and vague. We can define accountability as, "A personal choice to rise above one's circumstances and demonstrate the ownership necessary for achieving desired results—to *See It, Own It, Solve It, Do It.*"[3]

This work also requires a leader and a vision. Without these two critical components, excellence may be elusive and out of reach—not because it is hard and not because people do not desire to do their best. Excellence will be out of reach because a system requires an aim and a leader who will take intelligent risks. We need leaders who will consistently model the behaviors that will shape excellence. Leaders must understand that leading is about relationships and connecting to employees and stakeholders.

Are you ready to *do*? If you are, the chapters that follow will give you the foundation and practical actions to begin your journey. The actions are simple, but ironically, many leaders choose not to take these small steps.

Why? Perhaps, the small steps seem so insignificant to those looking for a big splash or instant gains. Creating a culture for excellence and transforming an organization is about building. And, as they say, "Rome wasn't built in a day." We cannot transform an organization to create a culture of excellence in a day.

Make a commitment to excellence

Whether we are talking about personal, group, or organizational excellence, one component is nonnegotiable—commitment. I am not a particularly athletic person, but at one point I had the opportunity to play in a local tennis league. My tennis experience consisted of what I learned in a physical education class in college—roughly six weeks of hitting against a backboard, lobbing the ball in the air, and several miserable attempts at playing the game.

I would not achieve my goal to compete in tennis if I did not make a commitment to learning skills, practicing the skills, and playing increasingly with more advanced players. I could never successfully compete in any tennis tournament if I only watched as others played tennis. I had to make a commitment and act to *do* what it took to improve my tennis skills.

So, I did just that. I practiced. I watched videos. I practiced. I bought a decent racket. I practiced. I played against opponents better than me and some who were worse than me. I made a firm commitment to excel in tennis. Now, I wish I could tell you that I went on to Wimbledon, but I did improve and continued to play against those who were better and more experienced. In a league that classified teams from F (least experienced) to A (awesome and experienced), I started on the F team. With commitment and perseverance, I transformed my tennis ability and worked my way up to the C team.

My tennis acumen may seem unimpressive, but for me, it represented quite an accomplishment. Success, no matter how small, did not come by blaming others for my poor performance. I committed to the basics to continually improve my serves, my agility, or my understanding and execution of good form. I never once said I just wanted to be okay at playing tennis. I wanted to be my best and made a commitment to do just that.

I often wonder why organizations choose to remain mediocre but claim excellence as their vision. Leaders have told me that using tools such as Six Sigma, ISO, process management, etc. detracts from the real work of the company. They lament, "We just can't take the time for that in this economy and competitive market."

My response? You cannot NOT take time to do the things that will build your competitive advantage, create a culture of learning and excellence, develop loyal customers, and actively engage your workforce in accomplishing your vision.

In the eyes of the customer, whether you sell shoes, run a nonprofit, manufacture high tech gadgets, treat illnesses, or educate our children, being "sort of" good is simply not enough. Buyers, patients, students, parents, teachers, and communities want our best.

Yet, many of us work exceptionally hard at just okay. What disturbs me most is that we often do not even realize we made that choice.

We make excuses, we blame the economy, the laws, the competition, and whatever or whomever we believe has created an obstacle to our success.

Performance excellence, by organizations, individuals, or teams, rarely succeeds without a clear vision of the outcome and the commitment and perseverance to stay the course. Attitude, a way of thinking, an environment, and unwavering support pave the road to success.

So, how much time does it take to build a culture of excellence? You will not like the answer: "It depends." Unless you take the first step and unless you start with some inkling of a path or guide, nothing will change. Creating and nurturing a culture is never a linear process with clean lines and directions to the next step. The work often feels messy and disjointed. Thinking about excellence is not doing. Just start.

Adopt a culture of excellence model

Do you need a place to begin thinking and planning for excellence? Consider three critical components that can serve as a roadmap and compass for transformation.

While the components and actions appear linear, the work is recursive and ongoing. No matter where or when you choose to create a culture of excellence, it always, always, must begin with a commitment to excellence and the persistence to press on.

The framework I use to guide transformation comes from the work of Dr. Bryan Cole, professor at Texas A&M University and consultant. As I began working within and with organizations, the steps in Dr. Cole's "Building a High Performing Learning Community" grounded my thinking and the actions I facilitated as we began to take purposeful and intentional steps toward transformation.

The Culture of Excellence Model, based on Dr. Cole's work, Figure 1.1, draws on W. Edwards Deming's System of Profound Knowledge, Peter Senge's core disciplines for building a learning organization, and embeds the Baldrige Excellence Framework and Criteria. Three interrelated core actions begin to shape and define an organization whose leaders strive to transform the way in which they lead, serve their customers, develop work systems and processes, and engage the workforce.

These concepts and actions are not difficult, but remember, you will not find a silver bullet or magical path to transform your organization. The processes, tools, and actions outlined in the pages ahead are simple by design. However, do not be misled into thinking they are unimportant or insignificant. Often, the most unassuming steps have tremendous impact, especially when consistently and collectively executed day-by-day, month-by month, and year-by-year. We build our path to excellence by one action and one process at a time.

Develop leadership and set direction

- Understand the system
- Lead for excellence
- Know clearly who you serve
- Create a pathway for excellence

Align the organization

- Value people and their contributions
- Identify key work processes
- Design a measurement system to monitor performance
- Establish a methodology to improve

Sustain excellence

- Embed continuous improvement in every facet of the organization
- Communicate that change is collaborative and an outcome of excellence
- Hold everyone accountable for excellence
- Celebrate excellence

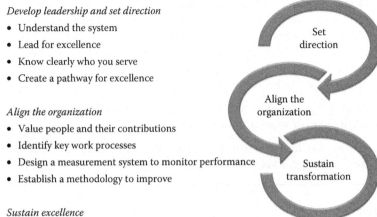

Figure 1.1 Culture of excellence model. (Reprinted with permission from Bryan R. Cole, EdD.)

The premise of this book is to help you build a culture of excellence one process at a time. Chapter two builds a foundation that is grounded in systems and systems thinking. Chapters three through ten explain each component of the Culture of Excellence Model. This model serves as your roadmap with specific processes and tools to begin to *do* and take the first steps toward excellence. Sharing this path to excellence enlightens, encourages, and supports all of us who have chosen to create great workplaces. Chapter 11 and the "Insights for the Journey" section of each chapter offer insights from those on the journey. These individuals bring their unique expertise and experiences from across a variety of organizations. These champions for excellence chose to share their learning with you, the reader. They have made excellence a way of life personally and professionally. Aristotle said it best: "We are what we repeatedly do. Excellence, then, is not an act but a habit."

Insights for the journey

Jeff Goldhorn: Ultimately the journey can only serve to improve your business model and to make you and your team stronger. Be careful that you keep the main thing, the main thing— Continuous Improvement and Excellence.[4]

Debra L. Kosarek: Others have asked me if I've ever felt "stressed out" by advocating continuous improvement. From their perspective, it is just a question of never being satisfied with the status quo. I look at it from a very different perspective. I know that if I'm engaged in continually improving I always have an opportunity to improve.[5]

Laura Longmire: Once you have been down this road, no leader wants to ever work any other way.[6]

C. Ryan Oakley: Be willing to listen. Be willing to learn. Be willing to change. Everything starts with you.[7]

Notes

1. J. W. von Goethe, *The Maxims and Reflections of Goethe*, trans. T. B. Saunders (New York: Macmillan and Company, 1893), 130.
2. H. Paul, J. Britt, and E. Jent, *Who Kidnapped Excellence? What Stops Us from Giving and Being Our Best?* (San Francisco: Barrett-Kohler Publishers, 2014), 92–97.
3. R. Conners, T. Smith, and C. Hickman, *The Oz Principle: Getting Results Through Individual and Organizational Accountability* (New York: Penguin Group, 2004), 47.
4. J. Goldhorn, PhD, interview by author, February 10, 2017.
5. D. L. Kosarek, interview by author, March 4, 2017.
6. L. Longmire, interview by author, March 2, 2017.
7. C. R. Oakley, DDS, FAGD, interview by author, February 3, 2017.

Set organizational direction for excellence

So, what is stopping you from your choosing the next step toward excellence?[1]

Harry Paul, John Britt, and Ed Jent

Transformation does not happen without a bit of grit. Setting the organizational direction for excellence to transform your organization begins with a clear understanding of systems and your specific system. Leading for excellence may require some tenacity and a willingness to learn new ideas and habits. We need to know whom we serve and what they expect from our products or services. None of this will happen if leaders do not create a pathway, a plan of action, for the future and excellence.

John Conyers, in *Charting the Course*, poses several questions "to test your own level of will and courage"[2] to leading and improving with an understanding of systems and systems thinking.

- Are you ready to confront the possibility that the systems currently in place in the organization, perhaps some that you designed, are creating mediocre results or not achieving the organization's true potential?
- Are you ready to address the issue that to achieve the kind of organizational excellence you aspire to, you may not have the right people in leadership positions, even if those people have become professional friends?
- Do you truly understand that creating organizational excellence is not a popularity contest but is being an advocate for the least

politically powerful subgroups of students? (or patients, line-workers, etc.)
- Do you have the will to impose greater responsibility on staff and greater accountability on the organization, even though you know you may meet with considerable resistance?

Notes

1. H. Paul, J. Britt, and E. Jent. *Who Kidnapped Excellence? What Stops Us from Giving and Being Our Best?* (San Francisco: Barrett-Kohler Publishers, 2014), 112.
2. J. G. Conyers and R. Ewy, *Charting Your Course: Lessons Learned during the Journey toward Performance Excellence* (Milwaukee: American Society for Quality Press, 2004), 5–6.

chapter two

Understand the system

The term "systems thinking" and the concept of "systems" is nothing new. We toss the words around and flaunt our intentions to improve the system, but do we see the interrelationships, the interactions, and how all the pieces fit together or clash against each other?

I suspect, based on the actions of many leaders, that these terms have only superficial meaning. Saying the words does not always imply understanding or that our actions reflect leading from a systems perspective.

This may be an unfair observation, but day after day, I see the problems created and exacerbated when leaders, or any of us, simply do not recognize or know how to respond to events as "systems thinkers." Although some people appear more adept at seeing the whole and the interrelationship of parts, we can all learn how to become system thinkers.

Transformation requires seeing with new eyes. This chapter sets the foundation for understanding and seeing the system.

What's in the room?

"List everything you see in the room." That began our study of systems. We listed things, many things, we saw in the university classroom. We created individual lists with limited references to relationships or people. We shared our list of things. A couple of astute classmates identified minute details that many of us had difficulty finding. We did not or could not see what they had seen.

Then, we began to categorize the disparate observations and look for relationships. We opened our minds to view what we had seen from new perspectives. The systems within the confines of that room and those that allowed us to be in the room took shape. We discovered systems that existed independent of us but still related to us. Until we took a deeper look, made connections, and asked questions, we were blind to the myriad of systems of a single room.

The exercise dramatically brought into focus the fact that we unconsciously look first at "things." This simple activity highlighted why we have difficulty in "seeing" the world as a system. If our view of the world consists only of linear and isolated observations, how much more difficult it becomes to see beyond the narrow focus of our lives. The exercise taught me that we tend to look for immediate responses. Instead, we need to look for not just symptoms or things but at the system in place.

What is a system?

The words system and process are very different concepts. Processes can emerge with or without a system in place. Deming defines a system as "a network of interdependent components that work together to try to accomplish the aim of the system."[1] Organizations conduct business every day without clarity of purpose or a vision. People work unaware of the interrelationships of their work. We fail to see how what we do affects many other components or processes. Along with our colleagues, we merely react to changing demands and external circumstances to remain competitive or even survive.

We tend to assume that if a company, an organization, or a school exists, then we have a system. Nothing could be further from the reality in most cases. We ignore the three critical implications inherent in the definition of a system:

1. A system must have an aim, a purpose, an objective
2. A system operates through intentional organization, interrelation-ships, and interdependence
3. The aim of the organization or whole has a higher priority than any of the parts

Many organizations have a written vision mission, or aim, which may be posted on walls, in memos, on business cards, and even t-shirts. Can we say we operate as a system because we have this physical evidence of an aim?

An aim without the network of interdependent components does not create a system. Interdependent components that do not work *together* do not create a system. Fancy displays and organization-wide rallies to pro-mote an aim waste time and resources if we ignore the second and third implications.

What are the characteristics of a system?

How do you recognize a system? Systems possess specific characteristics and elements that help us know if a system exists. Or, do we merely have a conglomeration of departments and people who may or may not be working together, interdependently, to accomplish the aim? Figure 2.1 illustrates the aim as the core of a system. An integrated system ensures that organization, interaction, and interdependence remain anchored to the aim. Look for the following:

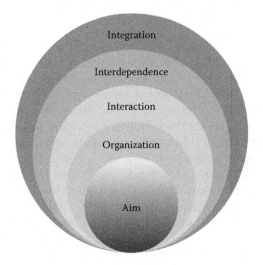

Figure 2.1 Characteristics of a system.

Aim

The aim of a system must be clear, create the outcomes required to achieve the aim, and consistently communicated. How do you communicate and demonstrate the value of the aim?

Organization

Every organization is designed to get the results they are getting. How are you organized? What are the reporting structures? Is your organization designed to support working together?

Interaction

The way we organize the system can affect the level and extent of interactions across the organization. If the "system" is not organized physically or culturally to provide opportunities for productive collaboration and dialogue, then those interactions may occur by chance, if at all. For example, how do you conduct meetings? How do you support collaboration and the sharing of ideas across work units and teams?

Interdependence

Departments, schools, work groups, team members depend on collaborative work and each other. Specific and measurable plans coordinate and link work efforts to maintain focus to get things done.

Integration

The plans, processes, actions, and decisions of a system support the aim of the organization. Integration within a system involves more than alignment of efforts—all the individual components operate as a whole. That means that the good of the whole in achieving the aim guides the work of subsystems, processes, and people within the organization.

One way to visualize the difference between a reacting, aligned, or integrated system is to think of departments, subsystems, processes, or programs as arrows as shown in Figure 2.2. Without a well-defined, clear, and consistently communicated aim, we tend to react to events, to others, and to external forces. If you work in this type of organization, you are likely to feel the tension of cross-purposes and lack of direction daily. The right hand does not know what the left hand is doing, or worse, we duplicate efforts and create conflicts with unintended consequences. We often do not know the source of this discomfort, which leads to blaming and further disconnection of work units.

With a clearly communicated aim, we have an inherent compass that drives decisions and begins to align the parts into a whole—into a system. Alignment occurs when plans, projects, and decisions all support achieving the aim. Do all the arrows point us toward the aim?

A high performing system moves beyond alignment to integrating processes, plans, and decisions represented by the arrows, working holistically to achieve the aim of the organization. Integration results when the components of the system work in tandem, where the arrows not only align but also support and complement each other. Effective integration creates the tipping point that differentiates good organizations from those that realize performance excellence.

The human body provides the best example of integration. The organs and components of the body must work together to keep us alive. When the heart fails to pump enough oxygen or the brain stops communicating, the body goes into stress or ceases to function. A healthy human body demonstrates integration at its best, when every function, or process, works together seamlessly and without conflicts or silos.

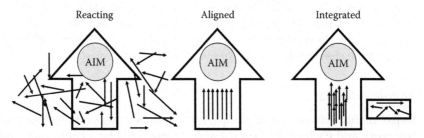

Figure 2.2 Reacting, aligned, and integrated systems.

Moving toward an integrated system also means taking things out, abandoning, or deferring good projects and programs. The arrows in the small rectangular box in Figure 2.2 represent those programs that do not support the aim. In an integrated system, we look at how every facet of our work ultimately moves us closer to our goal or aim. Anything that veers us off course, even very good things, may not be right for right now. Be forewarned; the stuff in that box belongs to someone, and they will not let go easily.

This is the point where many organizations falter. Do we have the discipline and methodologies to know if the work within the system is on track, and what do we do if it is not? Integration becomes a bit harder to see and understand. Think of the human body where every function supports the aim of "living" and consider how your organization:

- Designs processes that support what is important to the organization and what it aims to accomplish both now and in the future
- Measures the extent to which processes work as designed within subunits and across the system
- Develops plans, analyzes performance, makes decisions, and ensures that the aim of the system is not hindered by silos and competition between and among work groups or processes

What are the key elements in a system?

While the characteristics guide our understanding of a system, several critical elements describe how the system works to achieve the aim and create value internally and externally.

Outputs and inputs

The starting point for the work of the system always begins with identifying the need, requirements, and expectations of the customers. Who will use what we produce, which may be a product, a service, or an outcome? The quality and value of what the system produces (outputs) depends on the quality and appropriateness of the input.

Processes

A process, which we will address in more detail in Chapter seven, includes the steps of taking inputs and transforming them into one or more outputs.

Control

Systems need a central control point to guide and manage the design, implementation, and evaluation of processes. We call this governance. Without guidance and direction through decision-making and planning, we lose our line of sight to monitor interactions, interdependence, and achieve integration.

Feedback

Improvement and sustained performance require mechanisms to feed information back to the system. Feedback can be positive or negative. You might think of it as our system thermostat. We either want to maintain a steady temperature, or we improve our comfort by adjusting the temperature. In a system, negative feedback sends up a red flag to alert us that we have deviated from the planned course and need to act to get us back on track. Negative feedback might also expose a lack of a shared vision or purpose. Positive feedback may signal that progress is steady—stay the course.

Environment

The environment consists of all the external elements that might affect the system. Sometimes, these elements determine how the system must function. For example, rules and regulations, the economy, or the physical location often constrain the system or influence the aim and how we accomplish work.

Boundaries

Boundaries define the system and establish the limits of components, processes, and interrelationships. These "fences" guide the system when it interfaces with other systems. Subsystems must also have boundaries that support building effective interactions and interdependencies across the organization.

A manufacturing company may create system boundaries on the type of widget to produce. A retail store may identify selling only organic foods; thus, establishing a boundary that influences wholesale purchasing and inventory. Hospitals serving only children have set a boundary for the type of patient they will treat. Subsystems within an organization often establish boundaries around roles and responsibilities or processes.

Who manages the system?

> When people in organizations focus only on their position, they have little sense of responsibility for the results produced when all positions interact... We do not see how our own actions extend beyond the boundary of that position.[2]

Peter Senge

The most serious consequence of not understanding or managing the system results when we do not take advantage of what the components offer to support the aim. This is suboptimization. We create blind spots in our estimation of the capabilities of the system. We miss the impact of challenges. We overlook the opportunities that circumstance, internal talent, or the environment can offer to provide customers our best work. Working this way breeds an adversarial environment, where people compete for limited resources. The culture becomes one of, "Get what you can before someone else does." In turn, the collaboration breaks down. We stop sharing ideas and resources. Unintentionally, we turn the organization in on itself—becoming our own worst enemy.

Management holds responsibility to focus the organization to accomplish the aim of the system. Leaders must bring clarity, so that everyone understands the aim and their role in moving the organization to that aim. Senior leaders "are the only people in the organization with a whole system perspective, positional power, and the resource control needed to determine both strategy and action."[3]

Deming[4] likened a well-optimized system to an orchestra, where the listening audience judges the quality of the performance and determines the value. Can you imagine the sound if every musician played different tempos, different notes, and competed for the attention of the audience? The conductor directs the orchestra and establishes cooperation among the players to achieve the aim of the orchestra as a system—beautiful music that meets the expectations of the listeners.

Contrast the orchestra with a bowling team, where the degree of interdependence is less critical to the outcome. While each bowler's score contributes to the team outcome, an individual bowler depends less on her teammates to achieve a high score; she must bowl well and obtain the highest score for the benefit of the whole team.

Most businesses from across sectors require a high degree of interdependence. As interdependence increases, so does the need for communication and collaboration. Just like the orchestra conductor, leaders oversee and provide the direction needed to ensure that the parts become a coherent whole—a system.

When leaders abdicate responsibility for managing the system, several consequences occur. They leave the aim and how to reach goals to chance. They hope that other senior leaders have the knowledge, understanding, and authority to facilitate or make decisions regarding the organization, interaction, interdependencies, and integration of the system.

Excellence requires a system and leadership that cannot be left to chance. Leaders who have not committed to change and leading the organization through that change become a detriment to the organization. Continuous improvement cannot exist if leaders fail to remain steadfast to their belief and commitment to excellence throughout the system.

How do leaders manage the system?

"A system cannot understand itself."[5] This conundrum creates a major challenge for leaders. Deming spent his professional career responding and researching a theory to guide leaders in how to understand the system to manage and lead transformation. He called it the System of Profound Knowledge. The theory provides the outside lens through which an organization can begin to see itself objectively. Figure 2.3 illustrates the four interrelated areas: appreciation for a system, knowledge of variation, theory of knowledge, and knowledge of psychology.

The intersection of appreciation of the system, knowledge of variation, and theory of psychology create the theory of knowledge. However, we must keep in mind that the components do not work in isolation. They are interrelated and interdependent. When we view them as separate and distinct, our view of the system remains fragmented. As we enlarge the center of the Venn diagram, which is Theory of Knowledge, we increase our ability to see and understand our system.

The System of Profound Knowledge does not require being an expert in each area. But it can form the structural foundation for learning how to lead a system. Without understanding systems, we will not fully comprehend how variation affects the whole. If we fail to consider the human dimension of the system—psychology of people, we will not fully

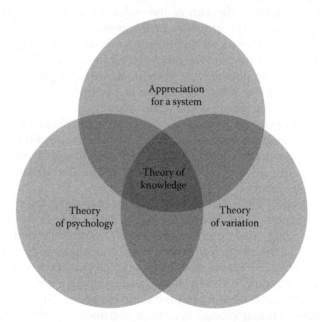

Figure 2.3 Interrelated components of the system of profound knowledge.

appreciate the processes at work in the system. Understanding the impact and causes of variation becomes more difficult. We need this knowledge to continuously improve the organization.

Appreciation for a system

Recall the definition of a system: "A network of interdependent components that work together to try to accomplish the aim of the system." The first obstacle is a clear aim. As simple as that sounds, few organizations move past the task of creating a mission or vision. This mission serves only as a rallying flag—without ever using the aim to drive decisions and give people a purpose for their work. Sometimes the misstep happens when we do not take the time to understand how the organization must interface within the broader environment, or we do not identify whom we serve.

Questions to ponder about appreciation for the system:

- Do you view your organization from a systems perspective understanding the parts and how they interrelate?
- Do you have a clear aim (mission or vision)?
- Have you communicated that aim across the organization and to your customers?
- How do you know everyone knows the aim or how it relates to their work?
- Do you make decisions and solve problems that support the aim?
- Do you consider how decisions impact the whole system?
- Do you ensure that one component (department, campus, subunit) does not gain resources at the expense of any other?
- Do you know the processes that support the aim?

Theory of variation

We cannot escape variation or eliminate it. "Variation is life. Variation will always be, between people, in outputs, in service, and in product. What is the variation trying to tell us about a process and about the people that work in it?"[6] Deming described two types of variations:

1. Common cause: This is variation built into the system. We know it exists. Hopefully, we know why it exists. We design the process to eliminate or mitigate as much variation as possible. In education, we know that schools serve children from prekindergarten through high school. Inherently, variation may occur in how we implement district-wide processes, communicate, or assess learning across grade levels.

2. Special cause: These are events that happen unexpectedly, were not previously observed, and may be unpredictable. They may occur as the result of changes within or outside the system. Changes to a process can generate unplanned consequences.

When we can predict the output and have determined how much variation can occur without affecting the quality of the product or service, we have a stable system. Far too often, we react to events. We implement solutions without a clear understanding of the predictable outputs of processes. We solve problems, where none exist. We tamper with the stable system and make things worse. The opposite response—over justification—causes us to ignore or overlook the opportunity to learn from special cause variation.

Figure 2.4 explains two common and harmful mistakes we make every day on how we respond to variation.[7] The first error attributes a mistake or variation to a special cause, when the result occurred as a function of the system and the processes within the system. The second mistake assigns a variation to the system (common cause), when the cause of the event was unique.

Ignoring the causes and source of variation makes continuous improvement more difficult and may produce flawed solutions. We can

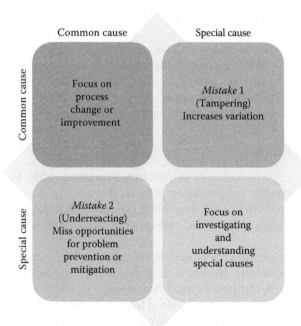

Figure 2.4 Common cause and special cause mistakes.

look at data all day, think we know what caused it but remain blinded to the cause, implications, or consequences. Deming explains it this way, "One may learn a lot about ice, but know very little about water."[8]

Questions to ponder about knowledge of variation:

- Do we know where variation exists in our organization?
- If we are not getting the results we expected, where do we look for answers?
- When we find variation, how do we tell the difference between what is expected (common cause), and what may be a "one off" or unexpected event (special cause)?
- How do we avoid tampering when responding to variation?
- What is our process for collecting, organizing, analyzing, and using data?
- How do we recognize a special cause?
- How do we respond to special cause events?
- What can or have we learned by investigating a special cause event?

Theory of knowledge

A key role of senior leaders is to make predictions—to have an idea, a theory of knowledge that guides decisions. First, let's look at a few definitions from the Merriam-Webster online dictionary. What is a theory? We can describe theory as "the analysis of a set of facts in their relation to one another." What is prediction? When we predict, we "declare or indicate in advance; foretell on the basis of observation, experience, or scientific reason." Finally, what is knowledge? Knowledge is "the fact or condition of knowing something with familiarity gained through experience or association."

Given those definitions, the role of senior leaders includes

- Using their existing knowledge and willingness to learn to analyze a set of facts about the system in their relation to one another
- Foretelling future performance based on observation, experience, or scientific reason

More simply, leaders are learners who understand the workings of the system and use this knowledge to guide the system in accomplishing the aim.

If we do not have a theory about what happened and how we got to that place, we cannot predict what will or should occur. Our predictions may lean toward guessing, but we have some solid information and understanding of the interrelationship within the system to improve and sustain performance.

The most disheartening circumstance occurs when a leader or a leadership team stops learning and does not value learning as a core competency of excellence. You have probably encountered people who do not even know what they do not know. Imagine an entire organization that operates from this position. A saying attributed to Daniel Boorstin and Stephen Hawking best captures this dilemma: "The greatest obstacle to discovery is not ignorance—it is the illusion of knowledge."

Questions to ponder about a theory of knowledge:

- How do we seek to understand the system and its interrelationships?
- How do we transform data into knowledge that informs problem solving and decision-making?
- How can we view the system from multiple perspectives (employees, customers, other organizations)?
- How do we analyze and make predictions based on what we know about variation and the components of the system?
- Do we have a process to improve?
- Are we aware of what we do not know?
- Do we ask questions?

Theory of psychology

Without grasping the interaction between people and the interaction of the system design and people, we would only see the organization as through a clouded window. We may see shapes but miss the details that bring what we see to life.

It is a fact: We are all different. Oh, how much easier it would be if we were more predictable. Perhaps motivation would be more effective because we would know how best to motivate one type of person rather than a myriad of experiences, backgrounds, perceptions, and needs.

External motivation does little more than promote compliance. Ranking people and their accomplishments may not build a dedicated workforce. Engagement and commitment to the organization and one's work only comes through intrinsic (internal) motivation. We must want to do our best work, and coercion rarely inspires our best work. In Chapter six, we will delve more into employee engagement.

Questions to ponder about a theory of psychology:

- How do we know whom we need to accomplish our aim?
- Do we have people in the right roles to leverage their knowledge and skills?
- How do we let people know we value more than their contribution and commitment? How do we value them as unique individuals?
- How do we communicate and build bridges of understanding across the organization?

- How do we know if these efforts are working?
- How do we support and design the system for team work and collaboration?
- Do we listen?

What creates results in a complex system?

> All organizations are perfectly designed to get the results they get—
> If you don't like the results you're getting, look at the organizational design![9]

> **David P. Hanna**

With Deming's System of Profound Knowledge as a foundation, we can explore more deeply the dynamics of how organizational capacity, system design, and human factors influence the intended outcomes or results of a system. Creating the conditions for a learning organization requires an understanding of the capacity of the system to achieve the desired results. Figures 2.5 through 2.8 will offer some insight into the challenges of managing complex systems regardless of sector or size.

Dr. Bryan Cole first introduced me to the concepts that follow. I vividly remember my "aha!" moment in his graduate class on quality management when he drew the diagram on the whiteboard. This revealed to me an entirely different perspective on what happens in our organizations. The visual explained the barriers, obstacles, and politics of the systems in which I worked. You may not be as profoundly influenced, but whenever I share these ideas in a class or at a conference at least one or two people react as I did. They remark, "So, this explains what happens and why we struggle with organizational improvements."

Note that in Figure 2.5 the outer oval represents the boundaries of the system. The system has an aim, a mission that defines its purpose with goals to help reach that aim. Two components influence the system's capacity to fulfil the aim: system design factors and human performance factors.

System design factors include the policies, subsystems, processes, and organization design. Both the physical plan and location comprise the infrastructure of the system. We can also view infrastructure from the perspective of the technologies that support the work of people in the system.

Human performance factors play a central role in how the system benefits from the knowledge, skills, and experience of people. Perhaps even more important are the attitudes and mental models that people possess. How we view each other and collaborate comes from our attitudes about

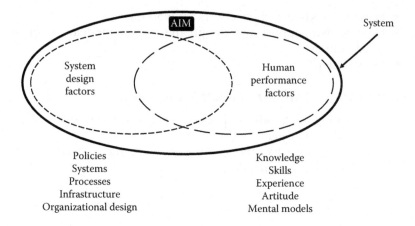

Figure 2.5 Organizational components: Managing the system for results. (Used with permission from Bryan R. Cole, EdD.)

ourselves and others. Mental models[10] consist of our assumptions, generalizations, our view of the world, and the lens through which we perceive and respond to everything around us.

People design systems, but systems also design people. The strength of the fit between the system design factors and the human performance factors impacts system processes and individuals' ability to perform. Figure 2.6 demonstrates how the meshing of system design and human performance factors begin to impact how leaders create policies, govern, build system, and create processes to reach the organization's aim.

Organizational culture emerges from the intersection of system design and people. We often think we can force culture, but forget that

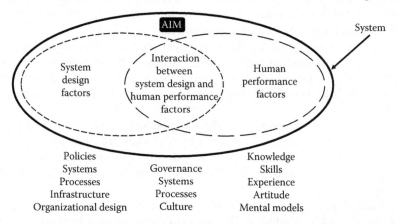

Figure 2.6 Integration of organizational components: Managing the system for results. (Used with permission from Bryan R. Cole, EdD.)

people *are* the culture. If everyone, including leaders, become intentional about what they want that culture to be, the culture will develop over time. You cannot dictate culture.[11]

Figure 2.7 adds another dimension to our understanding of managing the system for results. Learning is crucial to transforming an organization and performance excellence. As human beings, we instinctively seek to learn. It is in our DNA. We are born learners.

Humans constantly look at their reality and the vision of what they want or hope to become. This tension between now and the future pushes us toward learning.

Have you ever watched a child who sees something they want from across the room or out of reach? He will explore his surroundings and learn a way to get there and not give up until he does. Learning goes beyond acquiring information. The child may have a lot of information about the toy he sees on the top shelf. He must know more about how to get to that toy.

Learning how to get where we want to be or to achieve the results we want is "lifelong generative learning."[12] Learning organizations require lifelong learners—no exceptions.

The bold lines that segment organizational and individual learning help us visualize how organizational learning and the intersection between the two components comprise organizational learning. Organizational learning does not exist without individual learning.

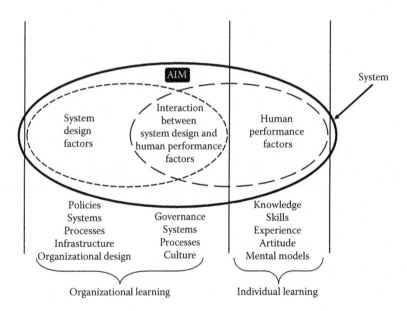

Figure 2.7 Organizational and individual learning: Managing the system for results. (Used with permission from Bryan R. Cole, EdD.)

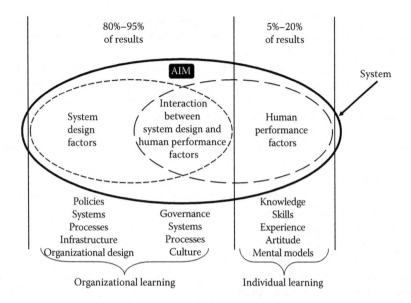

Figure 2.8 Managing the system for results. (Used with permission from Bryan R. Cole, EdD.)

Finally, Figure 2.8 displays a very sobering truth about the results in an organization—we can attribute only approximately 5%–20% of the improvement opportunities or problems on people, human performance factors. But 80%–95% of the problems occur due to system design factors and the intersection with human performance factors.

Throughout Deming's career, he asserted that around 95% of problems or mistakes belong to the system, and senior leaders (management) hold the primary responsibility for managing the system and the results. Special causes constitute the remainder of the mistakes. Clearly, blaming people will not lead to improvement. Do not assume that we must accept these mistakes. Rather, we must remain diligent, with leaders at the helm, to remember that the most effective way to improve and avoid future problems is to improve the system.

What is a systems thinking?

> Systems thinking is a discipline for seeing wholes.
> It is a framework for seeing interrelationships rather than things, for seeing patterns of change rather than static "snapshots."[13]

Peter Senge

Systems thinking is using the knowledge of the system to lead the organization to improve. Systems thinking becomes a way of being and approaching everything with a different lens. There is a caveat, and Donella Meadows [14] puts it bluntly, "For those who stake their identity on the role of omniscient conqueror, the uncertainty exposed by systems thinking is hard to take." What can we do?

We can do more than we think, but we must take the first steps. As a leader, as a systems thinker, openness and the willingness to look at your organization with new eyes requires strength to stand up against critics, persistence in ferreting out the root cause, and a keen eye for seeing the whole and all the interdependencies.

Before embarking on the systems thinking adventure, and especially as you invite others to join you, a few basic expectations may prepare you for the journey. Peter Senge[15] outlines what he refers to as the laws of systems thinking, which he identifies as the fifth discipline. Senge offers 11 laws, all of them important, but I suggest that you not venture out into the world of systems thinking without these three:

1. "Dividing the elephant in half does not produce two small elephants."[15] We cannot piecemeal systems thinking by looking at only one component at a time. Understanding the system pushes us to examine *all* the interdependencies that potentially affect the issue. Most important, this analysis cannot be stymied by organizational divisions or functions. We must remove the barriers and look across the boundaries that typically separate and isolate our work.
2. "Cause and effect are not closely related in time and space."[15] The cause of the problem you are facing today usually began long before the symptoms (effects) of the problem surfaced. Think sleuth—investigate, dig, and question. We assume the opposite of this law. Our starting point typically involves looking at whatever sits closest to the symptom or some person. Remember, we may not feel the pain (symptoms) for days, months, or even years from the initial cause.
3. "There is no blame."[15] There is no one to blame. We created the beast, and we must assume responsibility for the beast. I cannot overstress Senge's observation: "Systems thinking shows us that there is no outside: that you and the cause of your problems are part of a single system. The cure lies in your relationship with your 'enemy.'"[16]

Getting started with systems thinking

I started this chapter with a story about the exercise of seeing what was in the room. Do you want to see more than the *things in the room*? Along with the three laws, consider three first steps and a few questions to transform how you approach your work, your organization, and your life from a systems perspective.

Ask a few questions.

- How did we get to this issue?
- What happened prior?
- Has this occurred elsewhere?
- Do we know the underlying cause?

Identify feedback loops.

- What are the key processes associated with this issue?
- What are the handoffs, how are processes connecting, or not connecting across the organization?

Recognize mental models. Understand how mental models affect decision-making and individual and group perspectives.

- What are the key assumptions individuals are bringing to the situation?
- Are you seeking to understand or criticizing and blaming?

Strategies and tools to build capacity for systems thinking

Many different strategies, tools, simulations, and communication techniques can support your own understanding of systems thinking and your ability to guide your team in adopting a systems approach to continuous improvement. While examples and details for all of them lie beyond the scope of this book, the Appendix provides an overview of many of the most common and easily implemented tools and templates.

By far, "The Iceberg"[17] has evolved into my favorite tool for illustrating systems thinking and leveraging our knowledge of the system to affect change. As you know, most of an iceberg lies below the surface of the water. The visual of an iceberg guides us through a series of steps as we analyze and discuss a critical event or even a crisis. Figure 2.9 gives us a visual of an iceberg and each of the steps of our questioning.

Step 1: Events. At the surface, we see events. We observe what has happened and how we responded.

Step 2: Patterns and Trends. If we begin to explore just below the surface, we get a closer look at what has happened. We might discover that this same event has occurred not just once before, but many times. Data in a simple run chart can provide a history of the behavior or results of the event.

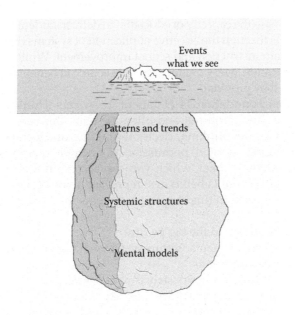

Figure 2.9 The Iceberg: Analyzing the system.

Step 3: *Systemic Structures*. Now we start serious exploration below the surface. Perhaps these patterns of behavior reflect the system itself. We soon realize that we have moved past just the single event and begin to realize the forces that have contributed to the event. How has our system design contributed to or exacerbated the issue? Remember—cause and effects are rarely close in space and time.

Step 4: *Mental Models*. Deep beneath the surface lurks our perceptions, our ways of thinking, the beliefs and values that people bring with them and the collective perceptions, beliefs, and values that have shaped the organization. How is our thinking or perceptions allowing the problem to form and persist?

Without digging to the bottom of this behemoth iceberg of our organization, we will continue to see only isolated events and miss everything below. The questions we ask during our search help us to put all the puzzle pieces in place. We begin to recognize the leverage points within the system that can affect deep change and improvement. A word of warning: Grabbing at easy solutions without exploring the entire iceberg may lead you back to where you began with the same or more serious problems.

We may need something very concrete and simple to understand a new or difficult concept. We need a tool that can spawn dialogue and deeper understanding. The Waters Foundation (http://watersfoundation.org) has a mission to increase the capacity of educators to deliver student academic and lifetime benefits through the effective application of systems thinking strategies in classroom instruction and school improvement. While their mission focuses on education, every organization or sector can benefit from the tools and resources they provide.

The Waters Foundation 14 Habits of a Systems Thinker[18] identifies the shift in thinking required and offers systems thinkers a flexible set of actions to support thinking, decision-making, and perspectives. The Waters Foundation granted permission to include their visual of the Habits of a System Thinker, which you can access on their website and see in Figure 2.10. These 14 habits also represent many of the concepts and suggestions discussed in this chapter. A systems thinker:

1. Seeks to understand the big picture
2. Observes how elements within systems change over time, generating patterns and trends
3. Recognizes that a system's structure generates its behavior
4. Identifies the circular nature of complex cause and effect relationships
5. Makes meaningful connections within and between systems
6. Changes perspectives to increase understanding
7. Surfaces and tests assumptions
8. Considers an issue fully and resists the urge to come to a quick conclusion
9. Considers how mental models affect current reality and the future
10. Uses understanding of system structure to identify possible leverage actions
11. Considers short-term, long-term and unintended consequences of actions
12. Pays attention to accumulations and their rates of change
13. Recognizes the impact of time delays when exploring cause and effect relationships
14. Checks results and changes actions if needed: "successive approximation"

What if?

So much more contributes to creating a culture of excellence through systems thinking, but everyone needs a place to start. What if you asked yourself or your team the questions in the introduction to Section I? How would you respond? How would your team answer the questions? Understanding systems and system thinking sets you on solid ground to begin to transform your organization one process at a time. If you accept the challenge and practice a few of the steps, you may be surprised at what you begin to "see in the room." *What do you see in your room?*

Figure 2.10 Habits of a systems thinker. (Reprinted with permission, Second Edition Copyright 2014, 2010 Systems Thinking in Schools, Waters Foundation, www.watersfoundation.org)

Insights for the journey

Debra L. Kosarek: Making "coming back to center" a routine, frequently performed activity. During strategic planning and strategy implementation, it has proven important to periodically evaluate whether the team is straying from our mission. It is easy to become so focused on innovation and improvement, the process takes on a life of its own. By periodically enforcing the discipline of bringing the discussion back to that of our mission, and what in the big picture we are working toward, we pull back from rabbit trails, and losing focus. As part of this activity, I ask, what do we want to have at the end of this process? Why are we doing this?[19]

Laura Longmire: We keep everyone focused through various types of communication. First our vision, mission and values are posted throughout all our sites, on our computer screens, and on all marketing publications. We open all meetings beginning with our vision, mission, and the core value of the month. The strategic objectives and balanced perspective are cascaded throughout the organization down to everyone's action plans. Quarterly reviews track the progress of these objectives and goals. All meetings have standard agendas aligned to our balanced perspectives.[20]

Anna Prow: In nonprofits, it's especially important to have a champion for the big picture, because the big picture is the organization's mission. All staff need to have shared understanding so that each team member can frame his or her own goals to advance that mission. I make the space so we all can explore our understanding of the big picture and reconcile the ways we may see it differently. I also foster conditions for executive leadership to share their big picture vision so staff can remain in synch with where leadership is taking the organization. I also place value on documenting the big picture by way of mission, vision, values, and goals statements. Structurally, I develop planning practices that connect the dots between the big picture all the way down to our individual work plans.[21]

Notes

1. J. G. Conyers and R. Ewy, *Charting Your Course*, 50.
2. P. M. Senge, *The Fifth Discipline: The Art and Practice of the Learning Organization* (New York: Doubleday Currency), 19.

3. J. G. Conyers and Robert Ewy, *Charting Your Course*, 3.

4. W. E. Deming, *The New Economics for Industry, Government, and Education*, 2nd ed. (Cambridge: Massachusetts Institute of Technology Center for Advanced Educational Services, 1994), 96.

5. Deming, *The New Economics*, 92.

6. Deming, *The New Economics*, 98.

7. W. E. Deming, *Out of the Crisis* (Cambridge: Massachusetts Institute of Technology Center for Advanced Educational Services, 1986), 318.

8. Deming, *The New Economics*, 101.

9. D. P. Hanna, *Designing Organizations for High Performance* (New York: Addison-Wesley Publishing Company, 1988), 36.

10. P. M. Senge, *The Fifth Discipline*, 8.

11. A. Cavanaugh, *Contagious Culture: Show Up, Set the Tone, and Intentionally Create an Organization That Thrives* (New York: McGraw Hill Education), 102.

12. Senge, *The Fifth Discipline*, 142.

13. Senge, *The Fifth Discipline*, 68.

14. D. Meadows, "Dancing with Systems" *The Donella Meadows Project Academy for Systems Change*, accessed November 15, 2013, http://donellameadows.org/archives/dancing-with-systems/.

15. Senge, *The Fifth Discipline*, 57–67.

16. Senge, *The Fifth Discipline*, 67.

17. P. Senge, et al., *Schools That Learn: A Fifth Discipline Fieldbook for Educators, Parents, and Everyone Who Cares about Education* (New York: Doubleday, 2000), 80–83.

18. "Habits of a Systems Thinker," Waters Foundation: Systems Thinking in Education, http://watersfoundation.org/systems-thinking/habits-of-a-systems-thinker/, accessed January 30, 2016.

19. D. L. Kosarek, interview by author, March 4, 2017.

20. L. Longmire, interview by author, March 2, 2017.

21. A. Prow, interview by author, February 15, 2017.

chapter three

Lead for excellence

> For many years, I have believed that organizational transformation—true deep and lasting change— will occur only when we call forth the leadership that resides within every person in the organization.[1]
>
> **Sister Mary Jean Ryan**

Everyone is a leader. No matter where you sit in the organization, you have influence on others. Some of us are leaders because we hold the top position in the organization. Others are leaders because they have the capacity to draw people to them. We can influence positively through our expertise or by our ability to have empathy and compassion. Many of us have a keen sense of what others need. We coach or mentor our colleagues. Influence can also breed contempt and apathy. Excellence only flourishes when we have positive influential leaders across the organization and at every level.

In this chapter, we will focus primarily on the role of senior leaders in the transformation of the organization and the leadership processes that support continuous improvement. This does not let the rest of us off the hook. Sustainable change depends on the commitment of everyone. We need leaders who take responsibility for the aim, a mission, for which we can strive. How would you, as a leader, answer these questions:

- How do you set the direction for your organization from a systems perspective?
- How do you model the values of your organization?
- How do you set the expectation for excellence through visionary leadership?
- How do you communicate?
- How do you establish a clear and understandable governance system?
- How do you support your community and contribute to your field?

I could add several more, but this should get us started. Notice anything about these six questions? They all include a common word, *how*. That little word and its cousin, *why* (the aim), can make the difference between mediocrity and excellence. "Perpetuating mediocrity is an inherently depressing process and drains much more energy out of the pool than it puts back in."[2] For the most part, we can all survive, but can we, are we willing to do our best? Do we know how?

What makes a leader?

> Example is not the main thing in influencing others,
> it is the only thing.

Albert Schweitzer

My view of leadership has evolved over time. I have soaked in the philosophies and perspectives of many people including, but certainly not limited to Peter Drucker, W. Edwards Deming, John Maxwell, Simon Sinek, Seth Godin, Stephen Covey, Jon Kotter, Jim Collins, Dr. Seuss, and all the leaders who crossed my path and modeled the essence of leadership. Read and learn from many teachers, but at the end of the day, form your own theory and create your unique version of you as a leader.

Following a conference presentation on leadership and engagement, a participant came up to me and said, "You gave a definition for engagement and you explained the role of leadership in engaging employees, but you never defined leadership. Leadership can mean different things to different people." I could not refute his reprimand. I assumed leadership meant the same to everyone.

Lest I repeat my mistake, how do I define leadership? You may agree or disagree, but I have learned that clarifying my intent might just keep you listening. I have a simple definition of a leader—someone I choose to follow.

The details come in what I expect of the people I consider leaders. These leaders may not have the position of CEO, but they inspire me to commit my very best to accomplish a goal—with excellence.

What do I look for in a leader? Let me count the ways. Leadership by any other name would still inspire (with apologies to William Shakespeare and Romeo[3]). We may vary on the most crucial characteristics of leaders, but I doubt that anyone could argue too forcefully against those identified in Figure 3.1.

We need leaders when the going gets tough, and the challenges become so burdensome that we cannot move forward. We get stuck in this cycle, repeating what we do and expecting something different to occur. Leaders create clarity, remove the barriers, and give us a picture of the future worth our efforts and commitment. "Real leadership is not about authority, control, or giving orders. It's not about titles or executive benefits. Leadership is about taking the initiative to do a job in a more efficient way or a better way, treating others with respect and compassion, and thinking of ways to be helpful. A leader is someone who is confident

Exemplifies integrity	• Clear about what they will not compromise
Communicates and clarifies the mission	• Makes the mission clear painting a vivid picture of the future so I understand why we do this work • Never says, "I'll know it when I see it." Well, if you can't see it, I can't help you achieve it.
Models personal and organizational values	• Personal and organizational beliefs and value are clear • Actions reflect those beliefs consistently and over time
Builds relationships	• Takes time to know me, our customers, and stakeholders
Demonstrates transparency	• Tells the truth—always
Initiates rather than reacts	• Persistent in the goal and resistant to criticism • Willing to challenge the status quo
Accepts accountability	• Takes responsibility for actions • Does not blame others, and can admit when she is wrong • Removes barriers to achieve goals

Figure 3.1 What I look for in a leader.

of her or his abilities and freely expresses that confidence—not in arrogance, but in humility."[4]

Organizations, even small ones, rarely have only one leader. The CEO, president, chancellor, superintendent depend on a team of leaders to facilitate and guide people in accomplishing the mission of the organization. Leadership teams have a responsibility to work collaboratively, communicate common messages consistently, model values, and share equally in their commitment to the mission of the organization. A cohesive leadership team serves as the driving force behind creating a culture of excellence. Unfortunately, "few organizations invest nearly enough time and energy in making leadership teams cohesive, and certainly not at the level of rigor that it requires and deserves."[5]

The best leaders form their team carefully. Individuals on a leadership team may disagree and hold different viewpoints. Strong teams do, but those teams never leave the room without knowing and supporting the decisions made and how to implement those decisions. Solidarity of purpose absolutely must exist to realize transformation of the entire system.

Can you reach excellence without a cohesive leadership team? Maybe, but the work will take longer, require more effort, and the risk to sustainability increases exponentially. Everyone, including employees, customers, and the community, watches how the leadership team demonstrates trust with each other. If they cannot see it or believe it, trust remains suspect.

When any one person on the leadership team walks out of the room and speaks disparagingly against another member of the team or the mission of the organization, they undermine the efforts of every person in the company or school. This does not mean that we should not speak out against unethical or illegal practices. It does mean that to improve, transform, lead for excellence requires a leadership team that shares a passion for the mission, trusts each other and those who do the work, and commits to working as a team for the good of the whole—the system.

What are key leadership processes?

Living in the world of processes and a system holds a level of complexity that many of us have difficulty accepting. Systems consist of processes, which means that everything we do affects everything else, directly or indirectly. This is messy work. This is nonnegotiable work in transforming an organization from mediocre, from, very good, to excellent.

The pitfall I see repeatedly—leaders give up their role as chief champion in the design and implementation of the key processes that can get us to the aim. The other slippery slope occurs when leaders assume that processes belong to everyone, except for them. Leaders who excel have processes that guide their work so that the organization benefits from their best work.

What are some of the key leadership processes, and what do they look like? Roll up your sleeves, we will embark on the journey to transform your organization one process at a time. The journey begins now. Figure 3.2 lists three critical processes and their importance. We will look at each one through the lens of a SIPOC. But first, we need to spend a little time building a common understanding of the SIPOC analysis tool and why we would want to take time using it.

What is a SIPOC?

Right, it does sound like some terrible disease. The SIPOC analysis tool provides a structured approach for analyzing a process. If you select only one tool for understanding the processes in your system, choose this one. Process maps or flow charts, fishbone diagrams, five whys, stakeholder analysis (see the Appendix), just to name a few, help you and your team dig deeper and provide even greater clarity on how a process works and *if* it works to meet the requirements of the customer. The SIPOC helps us to expose the context in which the process operates. It can also reveal

Leadership process	Why is it important?
Identify core values	Core values identify the beliefs and philosophy of the organization, support the vision, shape the culture, and reflect what a company values.
Conduct meetings	Meetings comprise a large portion of communication within an organization. Using this time effectively and efficiently increases the quality of decision-making. Effective communication builds trust, sets expectations, and supports engagement of employees, customers, and stakeholders.
Develop a succession plan	Succession planning identifies and develops future leaders by building internal capacity to sustain excellence.

Figure 3.2 Leadership processes.

handoffs and the importance of key inputs. Figure 3.3 gives an overview of the SIPOC analysis tool.

Variations of the SIPOC might add the requirements of the customer and the inputs. The order in which you ask each question can vary. You may find it easier to begin with customers and work backwards (COPIS). Some teams prefer to identify the high-level process steps first followed by customers, outputs, inputs and suppliers (PCOIS). For the purposes of this book, I have chosen to pose the questions for each process using the PCOIS order. The questions create a framework to increase the team's understanding of the process and the interdependence of the process with external as well as internal customers and suppliers.

Since the SIPOC is a tool for analyzing processes, we can use the SIPOC to better understand what the team needs to use this tool effectively. The SIPOC in Figure 3.4 shows who supplies information and what they provide (input). We can also identify the customers of the SIPOC and their expectations or requirements reflected in the outputs. Finally, the high-level steps add clarity and consistency for how we agree to create a SIPOC. Some of the steps may have subprocesses that we might want to put in a process map, but for now, we will keep it simple and focus only on the high-level steps.

Throughout the remainder of this book, the SIPOC questions help explain and clarify the key processes discussed in each chapter.

The best way to use these questions is to answer them from the context of your organization. The processes and the main five to seven steps can

SIPOC analysis tool

SIPOC purposes

- To provide an "at a glance" overview of a process
- To define the start and stop boundaries of a process (and project scope)
- To clarify relationships of the suppliers of inputs to the process
- To identify process customers (internal and external), and the process outputs that they seek
- To identify unintended wastes output by the process

The SIPOC analysis tool provides a big picture view of the important elements of the process. The team uses the SIPOC to gain a deeper understanding of the broader context or system in which the process occurs.

Before starting your analysis, clearly state the name of the process. Start with a verb. Define the boundaries, or the scope, of the process. Where does it start, and where does it end?

The order in which you ask each question can vary. You may find it easier to begin with customers and work backwards (COPIS). Some teams prefer to identify the high level process steps first (POCIS).

Suppliers	Who provides input for the process?
Inputs	What do they provide? What are your expectations for the quality of this input, such as on-time, complete, other product specifications?
Process	What are the 3–7 major steps of the process? State as a verb-noun.
Outputs	What is created by the process? What are the specifications for the output?
Customers	Who will use the outputs of the process? What are the expectations and needs of the customer for this product or service?

Figure 3.3 SIPOC analysis tool overview.

apply to any sector. What will change is how you respond to the questions. Who are your suppliers and customers? What are the specific inputs and outputs for your organization?

I have studied and designed many process across all sectors of business and education. In my work, I did not always use the term SIPOC, but we always analyzed the process using the thinking behind the tool. Sometimes, you must meet the audience half-way, which means not scaring them off with acronyms and strange sounding tool names.

My preference remains teaching these skills explicitly to build the capacity of teams and individuals to think differently about their work. I want them to understand how their work fits into the system. As we explore key processes for leaders and study the components of the Culture of Excellence Model, the SIPOC approach will illustrate concrete examples of the *how*.

Suppliers Who provides (supplies) input for the process?	• Members of the team working on the process • Other colleagues who either give info or use the output
Inputs What do they provide? What are your expectations for the quality of this input, such as on-time, complete, other product specifications?	• Information about the process • Identified process to analyze • Defined boundaries of the process • Information on who uses the process
Process What are the 3–7 major steps of the process? State as a verb-noun	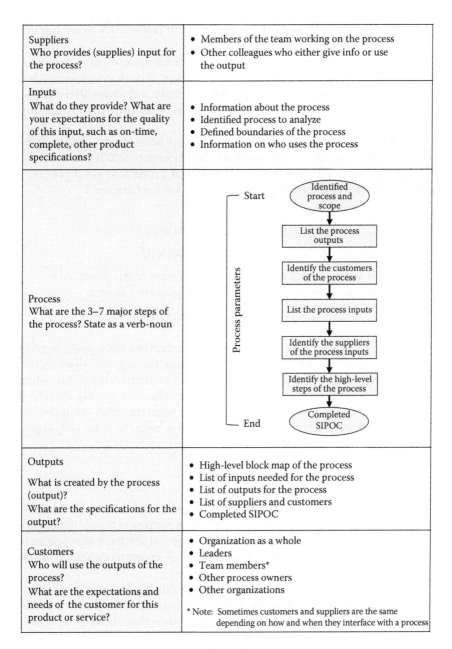
Outputs What is created by the process (output)? What are the specifications for the output?	• High-level block map of the process • List of inputs needed for the process • List of outputs for the process • List of suppliers and customers • Completed SIPOC
Customers Who will use the outputs of the process? What are the expectations and needs of the customer for this product or service?	• Organization as a whole • Leaders • Team members* • Other process owners • Other organizations * Note: Sometimes customers and suppliers are the same depending on how and when they interface with a process

Figure 3.4 Example for creating a SIPOC.

More than once, I have heard, "But this will never work in my organization. We don't have time for this."

Never lose sight of the context in which all systems and processes reside. The process may have different steps in different sequences, but I challenge you with this question: "Do you have a documented process that you and others can repeat, communicate, and improve to transform the organization?"

Without the consistency documented key processes provide, you have little evidence regarding your work or how to improve it. If you aspire to excellence, you need a systematic, intentional process to keep you on a path. Whether it is the right path is a discussion for Chapter 5.

How do leaders set direction, communicate, and monitor organizational performance?

> If your actions inspire others to dream more, learn
> more, do more and become more, you are a leader.

John Quincy Adams

Since processes interrelate and depend on each other for results, overlap will naturally occur. For example, part of setting the organizational direction includes your processes for planning and implementation. Modeling values begins with establishing the aim and values for the organization. In systems, nothing exists separate and apart from the other components. The three processes that follow can help leaders begin to establish, implement, and sustain excellence.

Identify core values

Core values describe an organization's "essential and enduring tenets—a small set of general guiding principles".[6] They represent a guiding anchor for behaviors and decision-making. The core values have the power to sustain organizations, push innovation, and create cohesive cultures focused squarely on the mission and goals.

Leaders must never enter lightly into establishing core values. When core values become nothing more than a token effort without any substance or modeling by leaders, everyone knows that *the emperor has no clothes*. Core values for decoration demoralize and disengage employees. If you stand long enough and listen carefully, any bystander can detect the difference between enduring tenets and words on paper—or the wall.

More than once, I have seen the rolling of eyes and smirks among staff when asked about the mission, vision, or values of their organization. The scenario unfolds this way.

"Tell me a little about the core values I saw posted in the front lobby. How do they influence or guide your work?"

"Well, I'm not sure what you are talking about, but if you mean the signs posted on the hallways all over the building, we never talk about those. No one pays much attention to them. We don't even know who decided on them. But we can tell you this, some of our leaders need a few lessons on those values."

Before you consider establishing your core values, ask this critical question first, "Are we willing to model these values through our decisions and behavior and make them nonnegotiable and embedded in everything we do?" If you have any hesitation or lack of commitment among key leaders, you risk losing an opportunity to build a foundation to stand up against fads, tackle change, or survive adverse market conditions.

Buffer, a company that offers social media and online marketing tools, discovered early in their growth the importance of core values. In the *Buffer Open* blog on October 27, 2015, Courtney Seiter explains how the founders began by asking a question, "What can give us the kind of feeling that makes work more than just ... work?" They believe that, "your values tell the world what you're about. They give your employees a reason for what they do—and your customers a reason to cheer for you."

As the Buffer team learned, establishing core values can also hurt. They saw and felt the pain of letting key leaders go once they solidified their core values. Leo, one of the Buffer founders commented, "That's what culture does. It's a disinfectant—it hurts a lot, but you end up being a lot stronger."[7]

I will not pretend that excellence comes easily or without pain. Making the commitment to hiring, operating, and sustaining excellence based on values and mission takes leaders who remain tenacious, fearless, and focused.

If you choose to take this journey, Figure 3.5 offers a simple, yet effective process for establishing core values. Who should participate? That depends. A small start-up company may begin with the founders. You might focus internally or expand the input to outside the organization. For example, a large school district may include a cross-section of staff, community, parents, and even students. Choose the approach that best meets your needs. What matters is what you *do* with your core values.

Conduct meetings

You can learn much about leaders and the organization by attending their meetings. Meetings consume a large part of every workday. What do we communicate to each other during those meetings? Meetings reflect the focus of the organization, the commitment to the mission and values, and set the tone for how this team will collaborate. Something as simple as an agenda that includes the mission, values, key goals grounds a team in the importance of their work.

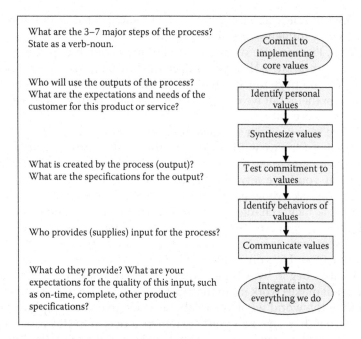

Figure 3.5 Process to establish core values.

How many hours per day, per week, per year do you waste in unproductive meetings? The number, if we choose to track it, would astound us, or worse, convince us that meetings are a worthless endeavor. How many of the following complaints describe your experience?

- Having meetings, just to have a meeting or to say, "we met"
- Meeting without a purpose or objective
- Not knowing *why* I am in the meeting
- Not staying on track or starting/ending on time
- Continuing to discuss the same topic and not reaching a decision
- Making decisions without enough information
- Not hearing from everyone—members withholding their thoughts until after the meeting

I imagine you could add a few more. Everyone has suffered through meetings that hijacked our day and zapped the life out of us. Whether we like it or not, business, our work, requires social interaction and communication with each other. Perhaps the problem lies not in the meeting, but our planning, conducting, and review of our meetings.

Meetings emerge as one of the first processes I encounter in my own and other organizations. How we orchestrate our time for collaboration makes the difference between a dreaded "meeting" and an opportunity to dialogue, challenge, and create great solutions.

We talk about two-way communication and develop elaborate plans for a variety of mediums to share our message. But we forget that two-way communication requires closing a feedback loop. It looks something like this:

- I tell you something.
- You listen.
- You respond to my "something."
- I hear your response and acknowledge that I understand.

Well-planned, effectively conducted, and reviewed meetings give us an opportunity to use two-way communication at its best. Many, if not all, of the complaints about meetings result from poor planning.

Do we know and invite only those who need to attend the meeting? Who can just receive information generated from the meeting? For some unexplainable reason, meeting organizers feel compelled to invite everyone—just in case.

Another major flaw in planning occurs when we do not develop a clear agenda that includes the purpose or objective for the meeting. What do we want to accomplish in this meeting? What actions will this meeting generate? Who will take ownership of those actions? How long will we discuss an item? At what point, do we defer a discussion because we do not have sufficient information to move to a decision or action? A good agenda answers those questions and prepares the team for what will or might occur during the meeting.

You may wonder why, of all the responsibilities of leaders, I chose this process. I have seen organizations weighed down and progress stymied because leaders do not know, or will not take the time, to conduct thoughtful and purpose-driven meetings. Communication by leaders to staff, customers, and other stakeholders begins with how they conduct meetings and use them to drive progress toward goals and the mission.

Bad meetings, over time, can erode confidence and harbor resentment. Effective meetings can bolster trust and contribute to positive relationships. As Joseph M. Juran[8] pointed out, "A good rule in organizational analysis is that no meeting of the minds is really reached until we talk of specific actions or decisions." Do your meetings result in specific actions and decisions?

I have included a sample meeting agenda template in the Appendix. You can find many variations of this template. Choose or create your own and use it. Watch what happens when people trust that you will begin and end on time, know why you invited them and what you plan to accomplish, and hold them accountable for the action items for which they are responsible.

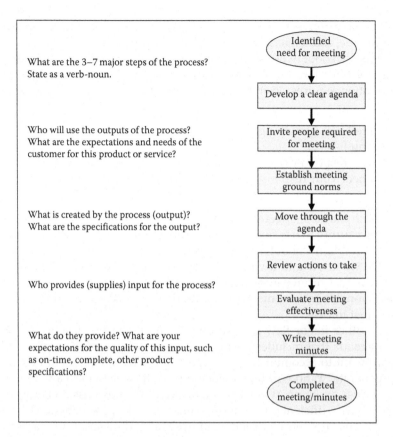

Figure 3.6 Process to conduct a meeting.

Figure 3.6 provides six critical steps for conducting meetings for an established need and closes with a brief evaluation of the effectiveness in the meeting. An evaluation can be as simple as using a Plus/Delta tool at the end of every meeting. I rarely close any meeting or presentation without asking, "What did we do well (plus)? What can I/we improve or do different as we work together (delta)? What action do I/we need to take to improve or prepare for the next meeting?"

As a final reminder, "Bring rigor and purpose to your organizational forums and meetings. If the water cooler or happy hour conversations are more honest than those in your meetings, you're not being effective."[9] Take a hard look at the format and substance of your next meeting.

Develop a succession plan

We talk about valuing employees and providing opportunities for advancement in our organization. Do we have a process that prepares for the future and offers opportunities to competent staff within the company? Do we have a plan for the future, and who will sustain effective performance of the organization, a department, or a team? We can apply the principles of succession planning to organizations of all sizes and sectors and any type of team.

Take sports as an example. In my house, we watch baseball, football, and basketball. We talk about baseball, football, and basketball. As someone who watches and listens between paragraphs of my book or playing scrabble with my online friends, I hear significant discussions around young players preparing for starting roles in the coming years.

I hear announcers speculate on a promising first year football player, or how a certain basketball player becomes a top performer. Coaches, at least winning coaches, recruit well and develop the skills and leadership capability of everyone on the team. No coach wants to face the coming year without the bench strength to sustain this year's success on the field or the court.

Why, then, do we fall into the trap of losing our brightest and best because they chose to move on? Why do we ignore the potential and capability of those we hired to lead our organizations in the future? Succession planning ensures that the right people with the right skills can pick up the ball and continue the work toward our mission.

The goal of succession planning includes having "every position critical to the success of the organization filled by internal candidates who have been carefully prepared to be successful in those roles."[10] It moves from reactionary to intentional development of the most important element of any organization—people.

Many organizations use the process in Figure 3.7, or some variation, to constantly prepare for the loss of an aging workforce, prevent the loss of institutional knowledge, develop the next generation of leaders, save time and money in recruiting and hiring, and sustain success. The uniqueness of your context or organization should never stand in the way of purposefully planning for the future.

What if?

What if you started with just one of these processes? Which one would you choose? Whichever you choose, keep this in mind: "...for an organization to remain healthy over time, its leaders must establish a few critical, non-bureaucratic systems to reinforce clarity in every process that involves people. Every policy, every program, every activity should be designed to remind employees what is really most important."[11]

The greatest leaders I know may not have the top position, but they all have one unwavering characteristic. They care deeply about others.

We are waiting for you; give us a reason and purpose to follow.

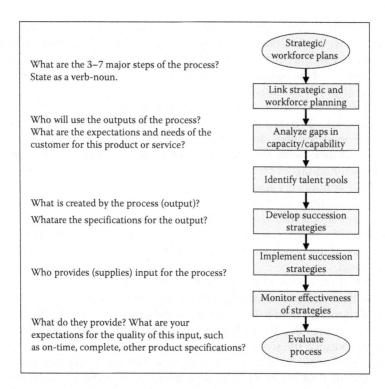

Figure 3.7 Process to develop a succession plan.

Insights for the journey

Ben Copeland: The process for improving and working toward excellence requires good leadership. I see it as steps that a good leader makes in an organization and understanding that these steps are processes. If you've got good leadership, you can transform things. If you are a great manager, you may transform a segment of the organization, but you may find it harder to impact the entire organization.[12]

Brian Francis: This is all about consistency in the tough moments. Stress kicks in when the agency is under attack by media, a licensee, or the legislature. People watch. Is he going to live with his core value of respect? Is he going to show integrity? Is he going to continue to be innovative when we are told to make cuts? That's when you are being watched.[13]

JoAnn Sternke: We model those organizational values in the decision making, and it is front and center on our agendas.

When I look at an administrative team agenda, it says first and foremost, "We need to develop our leadership skills, build management capacity, and work collaboratively to innovate our processes and systems all in the name of meeting our mission to open the door to each child's future." It is in our mind's eye. We keep in front of us that preferred future so that we are able to have our actions, behaviors, and decisions reflect those values inherent in our mission.[14]

Doug Waldorf: This comes down to organizational culture. Culture either enhances core values, or destroys those values. Recognizing and rewarding behavior that promotes and preserves your work culture is a key ingredient to lasting success. This begins at the top of the organization. Executive and senior leaders "living" the values each day will set a powerful and far-reaching tone permeating to frontline team members. Creating an environment where candid, direct two-way communication is easy to see and achieve is an excellent way to model core values.[15]

Notes

1. Sister M. J. Ryan, *On Becoming Exceptional: SSM Health Care's Journey to Baldrige and Beyond* (Milwaukee: ASQ Quality Press, 2007), 103.
2. J. J. Collins, *Good to Great: Why Some Companies Make the Leap... and Others Don't* (New York: HarperCollins Publishers, Inc., 2001), 208.
3. W. Shakespeare, *Romeo and Juliet,* Romeo to Juliet in Act II, Scene II.
4. Ryan, *On Becoming Exceptional,* 103.
5. P. Lencioni, *The Advantage: Why Organizational Health Trumps Everything Else in Business* (San Francisco: Jossey Bass, 2012), 20.
6. J. C. Collins and J. I. Porras, *Built to Last: Successful Habits of Visionary Companies* (New York: Harper Collins Publishers, 1997), 73.
7. C. Seiter, "The Untold Story of Buffer's Values: Why We Created Them and Why They Hurt", *Buffer Open Blog,* October 27, 2015, https://open.buffer.com/creating-values/.
8. B. Gaille, "11 Priceless Joseph M. Juran Quotes," accessed December 4, 2016, http://socialracemedia.com/11-priceless-joseph-m-juran-quotes-brandon-gaillecom/.
9. *The Baldrige Journey: A Practical Guide from Colorado's Experts,* e-book no date, accessed July 8, 2016, https://www.elevationscu.com/baldrige-ebook, 9.
10. M. Timms, *Succession Planning That Works: The Critical Path of Leadership Development* (Victoria, BC: Friesen Press, 2016), 2–13.
11. Lencioni, *The Advantage,* 16.
12. B. Copeland, interview by author, February 20, 2017.
13. B. Francis, interview by author, February 24, 2017.
14. Dr. J. A. Sternke, interview by author, February 22, 2017.
15. D. Waldorf, interview by author, March 13, 2017.

chapter four

Connect with customers and stakeholders

> A customer is the most important visitor on our premises, he is not dependent on us. We are dependent on him. He is not an interruption in our work. He is the purpose of it. He is not an outsider in our business. He is part of it. We are not doing him a favor by serving him. He is doing us a favor by giving us an opportunity to do so.
>
> **Mahatma Gandhi**

Whom do you serve? Do you know? Maybe, maybe not. I have participated in intense discussions where leaders or teams could not answer that question. At least, they could not agree on the answer to the question. Everyone had their own perspective on whom they served.

"Students are our primary customer," some argued.

"What about parents? They have expectations on the academic achievement of their children."

"I don't think we should forget about the community, or teachers, or our local industry."

Who won the argument? Consider the stated mission of many school districts. Although the wording and length differ, these two examples represent the focus for most educational organizations:

- We will prepare our students to succeed in a global tomorrow.
- We provide all students a high-quality education in a safe and nurturing environment where each student demonstrates a spirit of respect, responsibility, and a commitment to academic excellence.

Based on these two missions, the student appears to be the customer. But are students the only customer? I do not propose to take sides on this argument, but I have heard similar discussions in healthcare, manufacturing, government, and small business.

What stands out as most important is answering, "Why do we do this work and for whom?" Your customers are the reason, perhaps the only reason, your organization exists.

Customers play a vital role in the transformation to excellence. Their needs define the products and services we offer. Their feedback often drives innovation and improvement. Their success becomes our success. The realization of that success depends on if, and how, we respond to some of these questions:

- Whom do you serve?
- How do you listen to those whom you serve?
- What do you know about them and their needs?
- How do you determine the products and services to meet those needs?
- How do customers access products/services or information about them?
- How do you seek and respond to feedback from customers, positive and negative?
- How do you know you have met and exceeded customer expectations?

Customer service is not a department. Everyone owns responsibility for how you serve with excellence. Every action, every interaction demonstrates a philosophy grounded in building relationships. Will your customers never forget how you made them feel?

What are the key characteristics of service excellence?

> The magic formula that successful businesses have discovered is to treat customers like guests and employees like people.[1]

> **Tom Peters**

Leadership

Excellence in serving customers starts with leaders and permeates the entire organization. The aim of the system guides every aspect of creating a culture of excellence, including customer service. Excellence in service does not begin with customers. It originates through employees. Leaders who respect and build relationships with staff establish the foundation for excellent service to thrive.

I expect great service whether I am at a supermarket, an airport, or my doctor. This link between leaders and staff played out one day while

I was shopping. As I walked into the store, I overheard the manager. His questions to the staff did not come out kindly, even if he was just wanting to explain what to do. Tone changes everything. In a loud voice, he barked. "This display won't work here. Why did you set it up this way? Can't you see that it will block the entrance and flow for customers. Do this over, now."

The manager saw me, and turned to me smiling, "We are so glad you came into today. I just took over as manager and want to meet your needs and ensure we have the products you need."

What a contrast! My first thought was, "I don't believe you." His welcome lost credibility with me. I wondered how the poor stocking clerk felt when he saw the sharp change in the manager's behavior.

In this supermarket, cashiers rarely look up to say hello. They continue side conversations, complain about their hours, or whine because their feet hurt. I seem invisible to them—an interruption to their bantering. If this scene is typical, it does not give me hope that the attitude of most of the employees will improve or that my shopping experience will ever meet, much less exceed my expectations.

Mission/vision of service excellence

I do not know the mission of this store, which is part of a larger chain. Nothing within the environment of the store suggests that mission. For 13 years, the store operated without any immediate competitors. I had few choices, and the store had a captive audience. What I do know is that I find understocked shelves and employees who may or may not assist me or complete my transaction with a smile.

When you are the only store, hospital, school, cleaners in town, the risk of complacency increases. A new supermarket opened a few months ago. They have a full parking lot. The atmosphere of that store hits you immediately, and you know that everyone from the manager to the custodian understands the mission. They see me. They listen to me.

Known worldwide for excellent guest service, the Ritz-Carleton strives to create unforgettable experiences. César Ritz embedded his belief into the operations and culture of the organization. "The customer is never wrong!"[2] The Ladies and Gentlemen who serve Ladies and Gentlemen know, live, exude the Ritz-Carlton Credo, Motto, and the Three Steps of Service.[3] In 1999, the company received the Malcolm Baldrige National Quality Award, but they were an icon in the industry long before this recognition.

I had the opportunity to attend a meeting at the Dallas Ritz-Carleton and experienced firsthand their legendary service. Everything reflected a long-standing culture of exquisite service and impeccable staff. The company had graciously agreed to share with a group of leaders their philosophies of customer service and hiring, developing, and valuing their Ladies and Gentlemen.

The focus on building that culture and designing the processes to support their level of service demonstrated that any organization can do what the Ritz-Carlton has accomplished. The speaker did not divulge any exclusive trade secrets. She simply spoke of the commitment and the will to stay the course and remain true to their founder's vision of making guests the center of their work.

As we were waiting for our car in front of the hotel, I casually commented to my colleague that I wished I had stopped at the gift shop for a bottle of water before starting our three-hour drive back to Houston. Before I even took a breath after that sentence, a Gentleman came up behind me with cold bottled water for me and for each of my colleagues. I never asked for anything, but overhearing my wish, he sprang into action. They say, seeing is believing. Even as a visitor to the property, I felt seen, important, and valued.

Listening and learning

Listening to our customers sounds easy. We revel in their positive comments and shudder at every criticism. We stop listening because we do not want to hear what they might say. I asked a school superintendent if he had engaged his community and Board in his innovative education plans. He said, "No. If I ask them, they will stop progress, and I will not be able to implement my plans." Clearly, the focus of his efforts centered more on his plans than on building relationships. Without a doubt, children and their education took precedence in his thinking and actions. He missed the importance of listening to his customers and stakeholders.

I have heard those same sentiments from leaders in manufacturing and service industries. When our goals and actions fulfil our dreams without consideration of the customer, we run the risk of good intentions that fail to meet the needs of those we serve. By refusing to listen, we run the risk of alienating our customers. That risk became a reality for the superintendent. He not only lost public trust, but worse, the children in that community lost opportunities to benefit from innovative strategies for learning.

When I put myself in the role of the customer, I do not care about your product or solution. I want to know how what you do or sell will solve my problem. If my sink is backed-up, can you clear the line? I do not care if you have the newest equipment. I just care that whatever you use will solve my plumbing issues because I have a large family gathering in two hours.

Offering excellence to our customers requires a deep understanding of what they want and need now and in the future. We have a responsibility to look beyond the immediate expectations. What problems may exist in the future? What changes in the environment, technology, or economy will require a new solution for a problem they do not know exists. Listening can often give us hints into that future. By listening, we may discover that our customers have innovative ideas that we never considered.

Some of those innovations spring up from their complaints. More than once, I have heard, "Have you thought of doing this? What if we tried this approach?" Not every complaint comes with positive suggestions. If you listen enough and take the time to understand concerns, you will find that people begin to trust that you care about their problems. With that trust, comes the openness to share ideas and partner with you to improve your product or service. If you ask, customers will tell you the truth.

A word of warning—"don't ask if you don't want the truth and don't ask if you have no intention of responding to the truth." Every time I fly or stay in a hotel, I receive an email with a link to a survey. Of course, I would always respond with what went well and, occasionally, offer suggestions for improvements. After responding to these surveys for a couple of years, I finally stopped responding, even to offer praise. Why? I never received any type of confirmation that they listened beyond the typical words when you complete the survey, "Thanks for sharing your experience. We appreciate your feedback."

I received one of those surveys from a hotel in a small town in Kentucky. The staff was extraordinarily friendly, but the rooms needed new furniture and numerous physical repairs. None of these issues caused me any undue hardship, but I maintain high standards for my "on the road" home. One more time, I gave my honest feedback highlighting the excellent service from the staff and the opportunities I saw to improve the overall conditions of their rooms. For the first, and only, time I received a very kind email wanting to better understand my comments and explaining that plans existed for all the problems I had noted. The manager went on to add how he hoped I would be pleased with the upgrades upon my next visit. This was not a form email from customerservice@somehotel.com. This email came directly from the manager of the hotel. I replied and thanked him for taking a personal interest in my concerns. If my work had brought me back to that town, you can bet I would have given them another opportunity to meet my expectations.

Another way to gauge customer experience—start listening to your colleagues. How do they talk about those they serve? What mental models, assumptions, perceptions, do they hold about your customers? I can only imagine how the cashiers talked in the lounge about the crazy lady who insisted that the sackers not smash her bread or grapes. Try this experiment. Sit in the lunch room or lounge of your organization for a few days and listen to the conversations.

I have done this, and you would be amazed at what you can learn about service excellence, workforce engagement, and the overall culture of the organization. How people talk about what they do and whom they serve tells you if leaders respect employees, employees respect one another, and if employees respect their customers.

Informal listening, surveys, focus groups, and other listening mechanisms open the door to building relationships. The key to their success lies in our openness to what those strategies reveal and our commitment to responding and, perhaps, doing something different.

Alignment: action, processes, results

Do you want to delight your customers? Give them more than they ever expected. That level of excellence in customer service requires an intentional alignment of your actions, processes, and results. This intention begins with understanding your system and the focus of leaders to communicate a clear mission. They persistently set high expectations for excellence. Leaders and everyone in the organization model the organizational values in all interactions. Iterative and continual improvement of documented and communicated processes supports the level of alignment that can achieve transformational results.

Zappos began as an online shoe store. Their growth has been phenomenal, and now consists of 10 companies under the Zappos umbrella. If you visit their website, you will notice the alignment of their mission, actions, processes, and results. Their motto, *Powered by Service*, sums up the overall beliefs and values regarding customer interactions, employee engagement, and organizational culture. Ten values, beginning with, *Deliver WOW for* Service, powers the Zappos culture, brand, and business strategies.[4]

Would they experience the same success without this laser focus on serving their customers and their employees? Probably not since this alignment of actions, processes, and results differentiates them from their competitors. Their key leaders took a deliberate approach for engaging customers and meeting their needs. Like the Ritz-Carleton and Zappos, the Walt Disney Company has a simple but powerful vision for their theme parks: "Create magical moments for guests of all ages."

The moment you walk onto the property of any hotel or park in the Disneyworld complex, you realize you have stepped into a magical place. The Walt Disney Company exemplifies the successful habits of visionary companies that Jim Collins and Jerry Porras described in *Built to Last*.[5] Every child, every adult receives service fit for a prince or princess. I know I felt that way from the moment I arrived—magical. Shouldn't every customer experience delight with magical service?

Do you have a dentist? How are you treated when you arrive and throughout the procedure? As someone who has accrued plenty of seat time in dentist offices, I know the difference between being another set of teeth and genuine interest in me and my oral health. I first met Dr. Ryan Oakley in 2004. After 13 years, I would never go anywhere else and willingly drive the 25 miles to his office. Many dentists have practices much closer to my home, but none compare to the service and care I receive with Dr. Oakley and his staff.

What makes the difference? The mission of Spring Creek Dentistry is "not just fixing teeth—it's about changing lives one smile at a time." Every action from the receptionist, to the dental hygienists, dental assistants, office staff, and the dentists reflect this commitment to serving me. From the moment I walk in, I am greeted with a smile, offered a bottle of water, and rarely wait more than 5 minutes after my scheduled appointment time.

I marvel with every visit at the processes in place to ensure a positive experience. This does not happen by chance. Dr. Oakley and his colleagues ensure that I experience excellence at every visit. His philosophy around customer service puts the patient first. "While many doctors want their patients to like them, I learned early in my career that it is not about me. Customer service is all about how the patient feels about their experience and how they are treated. It is never about us."[6]

For each of these organizations, aligning their actions to their mission, establishing and continually improving processes, and listening to their customers have yielded positive results and raving fans. When it comes to connecting with customers and serving them well, your "success ultimately depends on what you have contributed to the success of your customers."[7]

What is voice of the customer?

Ask and listen. Take the time to learn and understand the circumstances of your customers. Never assume, never take for granted what you might discover. Surveys will give you some insight, but digging a little deeper and taking the time to meet face-to-face may provide information that survey questions cannot reveal. You might consider using a simple tool (see the Appendix for the Voice of the Customer template) to capture several key questions, such as

- Who is the customer?
- What does the customer want from us?
- What is/are the issue(s) that may prevent us from meeting their needs?
- What are the critical customer requirements (if you don't do this, nothing else will matter)?
- What is the measurable expectation to their requirement?

Always begin with the voice of your customer. Who are they? What do they need? How can we best fill that need? And, how can we meet their expectations above and beyond what they expect. The Ritz-Carlton, Dr. Oakley, and others do just that.

Understanding and responding to the voice of the customer includes several processes. We must know whom we serve and their problems. We must take our understanding of their needs and identify the critical requirements that could solve their problems. We must continue to listen to understand if the product or service we designed meets their expectations and solves their problems. If our customers have concerns, we must respond, find the root cause of the concern and create action plans to improve service, products, or communication.

How do you connect with customers and stakeholders?

> Ask your customers to be part of the solution, and don't view them as part of the problem.[8]

Alan Weiss

Being intentional about how you treat those you serve does not require a large budget. You can serve customers with excellence for no more than what it would cost for you to offer mediocre service. You stand to lose much more with poor service. What is the price tag for lost sales, lost trust, and ultimately, your reputation? The same applies regardless of the type or size of organization.

Jeanne Bliss identified five customer leadership competencies since "For customer experience efforts to become valued and considered critical to driving growth they must rise above the fray of being defined as problem solving or chasing survey scores."[9] At the core of these competencies lies the importance of a proactive and systemic approach to recognizing that your business or organization exists because of your customers.

1 Manage and honor customers as assets
2 Align around experience
3 Build a customer listening path
4 Embed experience reliability and innovation
5 Lead one-company accountability, leadership and decision making

Organizations that demonstrate excellent customer service exhibit those five competencies. To what extent do you consider customers as an asset indispensable to your success? Are your processes and work aligned to meet the expectations and requirements of your customers? Too often, we design our work system for our convenience rather than to serve our customers.

Do you listen, really listen? To Jeanne's point, we give surveys at the end of service and assume this information tells us everything we need to know about the customer experience. Sure, we do learn something, but we have become fixated on surveys as the be-all and end-all of listening to customers. Many companies seldom listen across all the critical interactions during the lifecycle of the service they provide.

You know your customer requirements, but can you predict how those requirements might change in the short- or long-term? In education, we often see changes coming based on demographics or new regulations. How do you use this information for planning now, and in the future? Are you able to adapt quickly to changing circumstances or to feedback at various touchpoints with your customer?

Serving customers belongs to everyone in the organization, and senior leaders hold the responsibility to model and lead the way. When I asked Dr. Oakley how he built the service culture at Spring Creek Dentistry, his response reflected how and where excellence originates in any organization. "I model the way I want everyone in our office to treat and serve our patients. If I don't do it, they will never see it as important. It begins with me."[10]

The three processes that follow represent common approaches from companies that have chosen to be purposeful and intentional about how they serve their customers and build sustainable relationships.

Listen to customers and stakeholders

How do you know if your product or service meets the needs, expectations, or requirements of those you serve? Simple—ask them. As easy as that sounds, many organizations assume they know or believe they know what will best address the needs of customers. You might think that retailers and service companies would understand this better than many government agencies or education institutions. Based on my shopping experience, if they are listening, they either ignored the customer or did not think it mattered.

Take music as an example. Upbeat pop music or hip hop might appeal to tweens or the under 20 crowd. Not so much for baby boomers. While music might create a sense of belonging and connection to the customer, it can also drive away customers who may choose to try you again, but likely will not.

Another side of this story might look like this. I normally do not shop at a local video game store but needed a gift for my 13-year-old nephew. I did not enjoy the loud music. If you only listened to me, and adolescents were your primary customer, you might miss the mark on what your primary customer expects from their experience in your store. I might never come in again, but those who enjoyed the music, certainly will. Know your customer.

What do they look like? What problems do they face? Where do they live? Many organizations falter because they lack clarity about whom they serve. Government and education often have predetermined service areas, but they must also understand who lives in these areas and how what they do can help them. Small businesses can flounder or fold because they attempt to solve everyone's problem. The same issue exists for nonprofits. I have seen nonprofits diminish their impact in the community because they try to serve too many groups, or they have lost touch with the changing needs of the people they serve.

We cannot serve the universe. Whom do we serve? What problems (needs) do they face? Have we listened to them? How can we meet those needs through our product or service? Figure 4.1 offers a straightforward approach to begin with the end in mind (the customer).

Identify customer requirements for products/services

Developing a new product or service does not begin with what we conceive as an astounding solution. Before you even begin to design the

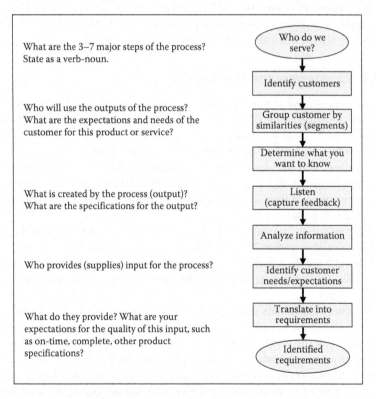

Figure 4.1 Process to listen to customers and stakeholders.

program, develop the product, or create a new service, start with a need. Have in your hand what you have learned as you asked and listened to your customer.

Implementing new programs in education often meets with harsh resistance. A group of bright, experienced, and caring curriculum specialists spent months on a new curriculum. The team knew the standards and considered the varied needs of teachers and the students they taught. As I began working with this team, I asked them what they had done to gather input from classroom teachers.

Their response? "We know what teachers want and what they need. We all taught or served as school administrators. We are very familiar with our students and our teachers." This group had a startling revelation after we assembled a group of teachers to listen to their perceptions of the proposed material and how it might impact teaching and learning.

Following an excellent presentation and overview of the new lessons and instructional support materials, they asked the teachers what they thought. After an uncomfortable silence, one teacher flatly stated, "I can't use any of this. Nothing that you have shown us will support the learning of my students. I am better off continuing to find my own resources."

You could feel the air get sucked out of the room. The look on the faces of the team said it all. They had, with great care and expertise, spent hours on this effort. How could this not be what teachers needed and could use? The team had to face the reality that once they left the classroom, they did not represent, nor could they assume, the role of classroom teacher. And, they simply failed to ask.

Once you have asked and clearly understood your customer, the process begins with a need—a need of your customer—as illustrated in Figure 4.2. The design or redesign of products, services, or programs does not begin until we know the specific requirements of what the customers consider essential to the solution of their problem or need.

Respond to customer complaints

Do you respond to customer concerns? What do you do about issues and concerns brought to your attention by customers? Listening to sales staff, receptionists, or staff who interact with customers reveals the answer to those questions. Staff, no matter their role or position, reflect the mind-set that exists within your business toward serving customers.

If we view complaints as a nuisance, they remain just that—a nuisance, an inconvenience in our day. Reacting to complaints might calm an angry customer. But what caused the problem in the first place?

Empowering frontline staff to handle most concerns, we can resolve issues immediately. When staff have the tools to effectively respond to

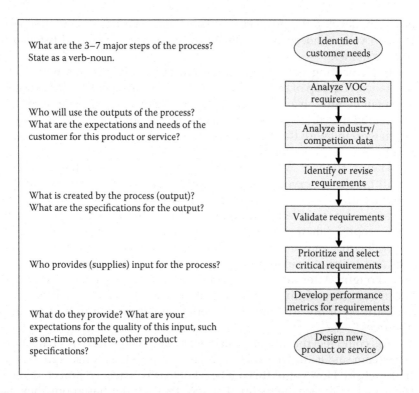

Figure 4.2 Process to determine critical customer requirements.

customer concerns and the ability to escalate a problem, we can address many issues at the point of service. In the example of a complaint management process in Figure 4.3, notice that the first process step states to "resolve complaint." The "resolve complaint" subprocess might include the following process steps for whomever encounters the customer:

1. Assess urgency of complaint
2. Listen
3. Apologize
4. Fix the problem or escalate if necessary
5. Thank customer
6. Follow-up with the customer

The immediate service recovery with the customer constitutes only one step in managing complaints. If we do not track the type or frequency of concerns, we will not know if this occurred once or twice or if customers experience this same issue every day. We can learn from complaints. Looking at them systematically helps us discover blind spots and opportunities to improve how we meet the needs of those we serve.

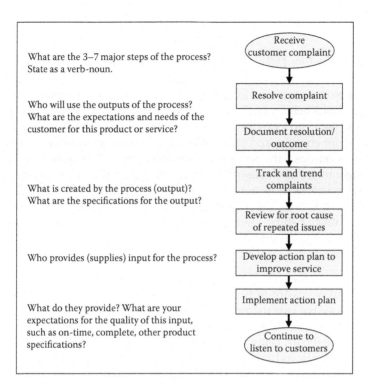

Figure 4.3 Process to respond to customer complaints.

What if?

Where should you begin to connect with those you serve? First, know who they are. Until you have clarity here, everything else will remain a best guess. You cannot serve everyone. You cannot be all things to all people. What if you started by simply asking and listening to the people who buy your products or services? Do you know their needs, expectations, problems? We think we know these answers. I have conducted enough focus groups to know that what we think they want and what we provide often do not align.

What if you watched, for one day, how employees and leaders treated customers? What would you learn? What if you watched the response of the customers? Do they walk away delighted or disgusted? Would they tell their friends and neighbors about how special they felt and how you offered a great solution to their specific problem? The advice of Charles Francis Adams, Sr., the son of John Quincy Adams and the grandson of John Adams, offers a good reminder, "No one ever attains very eminent success by simply doing what is required; it is the amount and excellence of what is over and above the required that determines the greatness of ultimate distinction."

Insights for the journey

Genie Wilson Dillon: High-performing leaders place importance on the voice of the customer and at every opportunity, this helps employees see that including the customer's voice, needs, and feedback helps with the alignment of actions, strategies, objectives, and data—to focus the work of the organization as well as the reason for changes and focus. Whether verbally or in writing—and in meetings—the focus on understanding and listening to the customers' needs should be a leadership priority. Some organizations use focus groups and community gatherings of customers to demonstrate the importance of hearing customers and building relationships. Giving feedback to customers after they give feedback to the organization or a leader is equally as important to demonstrate that their voice was heard and mattered.[11]

Debra L. Kosarek: Starting with the end in mind, we look at where we want to be in the next 5 years. Based on that destination, we can then create our milestones via our short-term goals. It is the most straightforward way for our team to follow a logical path tying budget and the strategic plan. While some organizations have a complex strategic planning process, I find it more important to keep things as straightforward as possible, for maximum staff engagement.[12]

Michél Patterson: If a company doesn't define itself based on the need it wants to fulfill for its customers, you are bound to be focused on your internal self. Someone focused on the customer is going to do it better. Our processes and strategy need to be focused on the customer experience. We must know how the customer is feeling and be very deliberate in defining that process to evoke certain emotions along the way. Our customer experience statement: "We want customers to feel trust, cared for, and pleased." Everyone in our organization from the president to the frontline staff wanted customers to feel trust that we will do what we said we would do.[13]

Notes

1. T. Peters. BrainyQuote.com, Xplore Inc, 2017. https://www.brainyquote.com/quotes/quotes/t/tompeters159515.html, accessed November 4, 2016.
2. J. A. Michelli, *The New Gold Standard: 5 Leadership Principles for Creating a Legendary Customer Experience Courtesy of the Ritz-Carlton Hotel Company* (New York: McGraw Hill, 2008), 1.
3. J. A. Michelli, *The New Gold Standard*, 23, 26, 31.

4. Zappos, "Zappos Family Core Values," accessed December 28, 2016, http://www.zappos.com/core-values.
5. J. C. Collins and J. I. Porras, *Built to Last: Successful Habits of Visionary Companies* (New York: Harper Business, 1994), 71.
6. C. R. Oakley, DDS, FAGD, interview by author, February 3, 2017.
7. P. Kotler, "Who Is Our Customer?," in *The Five Most Important Questions You Will Ever Ask About Your Organization*, eds P. F. Drucker et al. (San Francisco: Jossey Bass, 2008), 33.
8. A. Weiss, PhD, "Allow Customers to Be Part of the Solution," *Alan Weiss Blog*, February 10, 2017, https://www.alanweiss.com/hot-tips/allow-your-customers-to-be-part-of-the-solution/.
9. J. Bliss, *Chief Customer Officer 2.0: How to Build Your Customer-Driven Growth Engine* (Hoboken: John Wiley & Sons, 2015), 3.
10. C. R. Oakley, interview by author, February 3, 2017.
11. G. W. Dillon, interview by author, February 13, 2017.
12. D. L. Kosarek, interview by author, March 4, 2017.
13. M. Patterson, interview by author, February 28, 2017.

chapter five

Create a pathway for excellence

> What's the use of running if you are not on the right road?
>
> **German proverb**

If we are serious about transforming our organization, finding the right road will determine whether we stop at hope or move on to success. Creating a pathway requires a roadmap. That roadmap must account for sudden changes in our environment and include a method to know when we have veered off course. If we find that our actions have forced us down a different path, do we know how to get back to the road? Can we find an alternative that will bring us safely to our destination? Do we have a plan for excellence? If we have a plan, how does that plan address these questions?

- How will you determine where you are now and where you hope to go?
- How do you involve staff and stakeholders in creating the pathway?
- How do you identify your goals?
- Are the goals aligned to your mission, vision, and core values?
- How do you create action plans that accomplish your goals?
- How do you ensure that your budget supports your goals and action plans?
- How do you review the performance of action plans? How do you know they are working?
- If you have a plan to implement a new program, how do you know the program can address the need? How will you know if implementation has occurred with fidelity? How will you know if the program had an impact or solved the problem or supported improvement?

Why is planning important to organizational excellence and transformation?

> Leaders establish the vision for the future and set the strategy for getting there.[1]
>
> **John P. Kotter**

Leaders recognize the importance of planning and creating goals for success. Leadership teams may dismiss planning as a waste of time and something to check off their to do list. Having led many sessions of strategic planning, I have seen my share of folded arms, sighing, and furtive looks at phones—hoping someone or something could free them from this evil event. I always want to shout, "Wait! Don't turn your back on planning. We can create something great here."

Why do people hold such an aversion to strategic or any type of planning sessions? Too often, planning becomes an annual, compulsory activity that becomes *much ado about nothing*. So much effort devoted to so few results quickly becomes time wasted and a source of frustration because nothing changes. Like ill-fated New Year's resolutions, planning starts out with great intentions but lacks discipline to realize the projected outcome.

Great ideas and great plans find their way to paper every year in thousands of organizations. Those plans find their way on a shelf never to be seen again. One of the first questions I ask: "How does this work support your strategic or department improvement plan?" I want to know how this work aligns to what they hope to achieve. When the pause that follows becomes uncomfortable, I know that the team either does not know how the work relates to the organization's goals, has never seen the plan, or no plan exists. We, in effect, fly by the seat of our pants hoping that what we do achieves results.

Planning is a process, not an event. When we view the planning process as an event, we are tempted to cross it off the list, move on, and delegate the work to others. That approach creates several problems:

- No one accepts accountability for the actions required to implement the plan
- No one owns the results
- No one knows how to translate the actions into the daily work flow
- No one—or few—see any value in the plan or the process that created it

As a process, planning requires that each action plan has an owner. If no one takes ownership of the work of the plan, we lose accountability. The plan must have the flexibility to respond to changes in the environment. If new regulations, changes in the economy, or new information threaten the success of the plan, we must have the latitude to reassess our actions, revise as necessary, and move forward. Without a means to measure progress toward the expected outcomes of the plan, we have no yardstick of results to know if what we implement even works. Finally, the leader and the leadership team must commit to the planning process and hold others and themselves accountable for the results.

Although our plans may not roll out just as we designed them, the thinking and planning process remains important to closing the gap from our current state to the level of excellence for which we strive. The specificity of a focused plan gives us clarity. We know where we want to go. We know which road will get us there. We will know when we have arrived at our destination.

What is strategy?

> Leading for execution is not rocket science. It's very straightforward stuff. The main requirement is that you as a leader have to be deeply and passionately engaged in your organization and honest about its realities with others and yourself.[2]

Larry Bossidy and Ram Charan

Strategy is a plan of action designed to achieve a long-term or overall aim. Nothing complicated here—strategy enables us to make complex decisions and focuses our efforts. Focused and intentional planning gives us a roadmap to follow. It ensures that we know where we are going, why we want to get there, and how well we are doing in achieving our mission.

While many organizations write a plan, few leaders take the next steps. Not because the next steps are hard, but because the next steps require discipline and commitment to stay the course. Leaders face resistance. Strategy inherently creates change. We do not need a strategy or plan to maintain the status quo. We cannot achieve or sustain excellence if we plant our feet into the sand or worse, concrete.

Strategy aligns people, purpose, process, and plans. In Chapter 2, we discussed moving from reacting to an integrated system. Figure 5.1 illustrates how that might look relative to strategy by connecting the who,

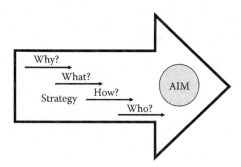

Figure 5.1 Strategy aligns the why? what? who? how?

what, how, and why. Who benefits from this strategy, and how does it add value? Who will carry out the plans? What is our mission? How does this strategy, these plans, move us toward that mission? Why does this matter? Can we commit to this work, this direction? Because quite frankly, if the strategy does not get us to our aim, we have taken the wrong path.

My personal and professional mission is and has been for some time to inspire excellence through people, purpose, and processes. My vision? Excellence, every day, everywhere, for everyone. If I take that mission and vision seriously, it means that I must plan each day with that as my primary, no, my only focus. I can never take my eyes off the road. I must look at what I do, what I say, how I go about my work with that as my compass.

If part of my personal excellence involves eating healthy and exercise, I must plan for that. I must measure my progress against my desired outcomes. I learned long ago that just writing down a goal got me nowhere. Creating a written plan at least started me thinking about what I needed to do. But not until I get up out of my chair, get on my bicycle, ride 10 miles, and commit to doing it regularly will I begin to see any changes in my stamina and strength. Without aligning my who, what, how, and why living healthy becomes nothing more than a wish.

A few premises regarding strategy will help build a context and connect several concepts that we have already discussed in previous chapters. Developing strategy:

- Uses a systems approach to change
- Creates a plan that manages strategic change initiatives
- Drives continuous improvement
- Focuses on critical initiatives that offer the greatest potential for accomplishing the mission
- Supports critical analyses of "what is" and innovative thinking for "what we can become"
- Aligns the work across the organization
- Creates a culture of cross collaboration that breaks down silos
- Holds everyone accountable for their roles and responsibilities
- Begins with strategic thinking

The visual of the three gears in Figure 5.2 illustrates the concept of first thinking strategically to understand the environment, challenges, and opportunities for excellence. From the insight gained by thinking strategically, we can develop a focused and visionary plan. The largest gear, strategic action, demonstrates the importance of effective implementation and monitoring of our plans. The extent to which we think, plan, and act on strategy determines our ability to integrate our work to reach and sustain excellence.

Figure 5.2 Strategic intent: thinking, planning, acting.

Strategic thinking

Thinking strategically gives us clarity about the present and a view of what might influence the path we choose to take. We must ask questions, lots of questions, if we hope to gain an in-depth understanding of the capacity of the organization, the external environment, and the level of commitment required to move forward. Strategic thinkers ask questions like these:

- What does excellence look like?
- What do we do currently and why?
- Why do we do things this way?
- What impact does this action have on everything else we do?
- What can we stop doing?
- Does what we do or want to do align to our mission?

Like the habits of systems thinkers, strategic thinkers develop certain habits and practices that guide their reflections and decisions. Strategic thinkers analyze, anticipate, examine multiple perspectives, and consider opposing opinions. They constantly look for ways to connect the dots for themselves and others. They embrace innovation, intelligent risk taking, and have an insatiable curiosity to understand the systems in which they live and work. A close friend reminded me often, "If you are not living on the edge, you are taking up too much room." I often resisted, but it pushed me to think beyond what I knew at that moment. I ventured closer to the edge to see what may have remained unseen. Strategic thinkers possess one critical habit—they remain learners.

Strategic planning

When you ask the questions, "What are your goals for the future, and how do you determine them?" Many will respond by explaining how they conduct this second component of strategy. Still others, especially in smaller organizations and businesses, do not see the relevance of planning and how this simple step can make a difference in their results. Planning involves understanding your reality and developing the actions to get you to your goal.

During an organizational assessment of a mid-sized company, I discovered that their planning process consisted of actions determined by the CEO, and each senior leader carried out those directives. Many of the initiatives included innovative solutions for meeting the needs of their customers. If the approach achieves results, why would we question the method to get there? Several problems surfaced as I interviewed the CEO's direct reports.

- Senior leaders or their staff had little commitment to the efforts. While the CEO had good ideas and intentions, she had not involved them in his private process.
- Senior leaders rarely expressed ideas of their own. They would defer to the CEO and explain what she wanted. They each had a list of actions for which they were responsible.
- Senior leaders rarely worked collaboratively on problems or issues.
- Everyone worked within their component with little regard for how any of their work connected or supported the organization and the mission.

I met with the CEO before finalizing my assessment and asked her about how they developed strategy. Without any hesitation and adamantly, she said, "I don't believe in wasting time with strategic planning. I have never seen the value. I can't tell you how many times I sat in multi-day meetings participating in silly activities."

"Then, how do you continuously improve your work to meet customer expectations and know what goals and actions you need to achieve your mission?"

"Simple," she said. "I thoroughly research innovative ideas and solutions for products and services and from that work to develop a list of initiatives. I give each senior leader a list of actions needed to implement those projects. This way, I don't need to go to the board for approval of some meaningless plan, and we don't waste time."

Probing for more information, I asked, "How has the board responded to your initiatives?"

She sighed and proceeded to explain how her Board of Directors had begun questioning and interfering with the implementation of several of the most promising projects. At this point, I asked, "How long do you plan to remain in this organization?"

"Maybe another three, but no more than five years."

I began thinking about my conversations with her direct reports who had never received any coaching or modeling on how to think strategically. They never had an opportunity to develop collaborative plans to reach their mutual goals or even understand their current condition as an organization. I saw few measures of progress related to any of their initiatives, so I doubted that they or the CEO knew the effectiveness of these initiatives.

My next question gave her something to think about. "When you leave, how will this group of leaders know what to do? How will they know how you made your decisions? Do you think they will have the leadership skills to lead this organization and develop a strategy that engages your stakeholders?" The silence that followed suggested she had not considered the impact on the development of her leaders or the sustainability of any of her best innovations.

A quick search of "strategic planning" on Amazon revealed nearly 25,000 books. Google gave me more than 103 million hits. It appears we have more than sufficient advice and models from which to choose.

No matter what method you choose, strategic planning in any shape or form, starts with knowing your mission. It defines the direction or goals, allocates resources, and specifies the actions to realize those goals. Leaders who keep it simple, attainable, realistic, and collaborative create plans that engage staff, customers, and stakeholders. Most important, they achieve results.

Strategic action

What we generally refer to as strategic planning stops short of comprehensive execution of the actions intended to deliver key outcomes and results to reach our goals. "Intentions are all fine and good, but it is the translation of those intentions into concrete items—mechanisms with teeth—that can make the difference between becoming a visionary company or forever remaining a wannabe."[3] Execution gives us the traction to move words on paper to actions and results.

What gets in the way of execution? The obstacles I see repeatedly across organizations have nothing to do with the quality of the plan or even the process used to develop the plans. We stumble for six main reasons:

1. Lack of focus on the critical few actions to achieve our goal
2. Failure to commit to following through and accountability

3. Absence of buy-in and communication with the people responsible for implementing the plans
4. Blame and rationalization of why implementation cannot occur
5. Discomfort and fear of change
6. Overloading human and financial resources—inability to abandon ineffective or inefficient processes, projects, or programs

Just as planning is a process and not an event, "Strategy execution is not a moment in time. It's thousands of moments across time."[4]

Leaders begin to lose confidence in the action plans when they do not see immediate results. We cannot expect results when we fail to identify the mechanisms to monitor the implementation and progress along the way. Effective execution of our plans depends on

- Explicit actions
- Specific timelines or reasonable expectations for implementation
- Clearly defined roles and responsibilities
- Accountability for the completion of the actions
- Measures to know if we are moving in the right direction
- Commitment of the resources required to accomplish the work

The problems associated with the strategic planning component of strategy rest here—execution. We may have invested time in strategic thinking and involved everyone in developing goals, but did not write specific action plans or include a way to review progress of those plans. If we do not move to this final stage, we become travelers who prepared for the trip, packed all their necessities, asked the neighbor to feed the cat, but never bought an airline ticket.

We stand bewildered, frustrated, and angry at the gate when they will not allow us to board without a ticket. "Execution is fundamental to strategy and has to shape it. No worthwhile strategy can be planned without taking into account the organization's ability to execute it."[5]

How do you create a pathway for excellence?

> When it is obvious that the goals cannot be reached, don't adjust the goals, adjust the action steps.
>
> **Confucius**

Develop strategy

Leadership teams must ask themselves, "How engaged and committed are we to this strategy? Will we all support the execution of the strategy?"

When leaders walk out of the room and criticize the actions that we have agreed will move us toward our goals, we immediately, and sometimes irrevocably, sabotage the planning process. Think of it as planting land mines along the road or removing the road signs leaving people confused and wounded. No wonder we become skeptical of how planning can drive improvements and excellence.

Figure 5.3 lays out a straightforward process to create your pathway to excellence—your strategy to achieve excellence. The process begins with strategic thought about your organizational challenges and priorities. A scan of your environment considers the economic, social, and regulatory

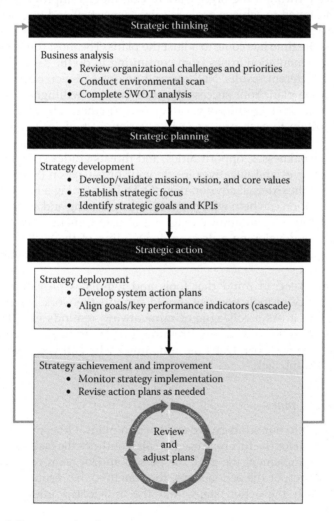

Figure 5.3 Process to develop strategy.

changes that may impact your ability to sustain improvements or reach higher levels of achievement. Understanding the context in which the organization operates allows leaders to consider how the system interacts with interrelated external systems.

The strategic planning component of this process includes developing or validating the mission, vision, and values. Clarity of purpose leads the way to defining the strategic focus and the goals for accomplishing that mission. Without indicators of success, the extent or level of progress remains a best guess.

Strategic action begins with specific action plans that align to goals and cascade through the organization. Continuous improvement of the actions designed to achieve goals requires consistent and ongoing monitoring of results. If results reveal that progress does not meet expected outcomes, the time has come to adjust the plans. The most common obstacle for implementation arises from continuing to do the same thing and expecting different results. Nothing and no one says you cannot make a change in the plan. Your success depends on your willingness to course correct based on the results of your key performance indicators.

In addition to each process step, a variety of tools or templates can support this work, such as the SWOT analysis. Consideration of your strengths, weaknesses, opportunities, and threats guides you in identifying the positives and negatives inside your organization (S-W) and outside of it, in the external environment (O-T). Developing a full awareness of your situation can help with both strategic planning and decision-making. Brainstorming, an affinity diagram, or a force-field analysis can all contribute to developing a deep understanding of the system. You can find overviews for these and other tools in Chapter nine and templates in the Appendix.

Always keep in mind that tools and templates provide structure, and the choice of tools you use should support the work, not become a hindrance to it. As a colleague of mine always reminds me, "Don't get wrapped around the axle." As I remind those with whom I work, "The tool police do not exist. Don't use a hammer when all you need to do is push in a thumb tack."

Write action plans

To give teeth to our strategy, we need action plans. Otherwise, our goals remain a wish for the future. An action plan outlines the tasks or activities required to implement the goal, the person and/or team responsible for implementation of the action steps, the timeline, the resources required to accomplish the action steps, and the anticipated internal/external collaborations.

Action plans need the collective thoughts and ideas that can only come from a team. When we collaborate and bring our expertise to the table, we strengthen the implementation and buy-in for the change this plan will bring.

Action plans must provide a means for measuring organizational performance. The monitoring and reporting process for each action step provides an effective system of accountability.

Action plans do not just occur at the organizational level. At every level of the organization, we create plans and measures that support the goal because everyone has a stake in the success of our strategy. When our entire organization sees how their piece, however small it may seem to them, aligns to our mission and goals, action plans become significant drivers of our success. This creates alignment.

When we write meaningful action plans, we have a tool to move the organization toward the achievement of goals and objectives. Think of an action plan like a project. It has a beginning and an end.

Action plans are not like the "never ending salad bowl" or "all you can eat" at your local diner. How often do you walk into a restaurant and order this way? "I need to eat. I don't have any money to pay for it. I want something tasty and filling, but I don't know how much I need so I don't leave hungry. And, by the way, I don't care how long it takes."

We would never order a meal this way. So, why would we approach filling a need for improvement (action plans) without resources, without knowing what to do, without a way to know if what you did worked, or how long it would take? If we want to create a pathway to excellence, writing down specific actions and monitoring their implementation become the most critical first steps.

The process steps to write an action plan depicted in Figure 5.4 can guide the development of plans at any level of the organization including divisions, departments, and even individual improvement plans. The key to the success of this process lies in keeping the action purpose statements succinct. The Appendix includes a template for writing an action plan and some additional explanation.

Review the strategy

Reviewing our strategy requires "robust dialogue to surface the realities of the business."[6] We must hold everyone accountable for the results and conduct open discussions. Consistent follow-through ensures that plans remain on track, and if not, determine the root cause and adjust the plans. The CEO referenced earlier in this chapter never allowed open discussion about the strategy, and she missed valuable input from her capable and experienced senior leaders.

Determine the team members who will be the primary individuals responsible for writing, implementing, monitoring, and or reporting on the strategy action plan.

1. Review the strategy and agree on the intent and purpose of the strategy to achieve the goals. What do we need to do to achieve the strategy?

2. Develop KPIs: How will we know that we are accomplishing that strategy? What measure of performance will we monitor to determine the progress?

3. Identify the initiative, projects, or programs to achieve the strategy. Write an Action Plan Result Statement. A specific result statement reflects the outcome of a series of actions needed to implement and accomplish a Strategy. A good Action Plan result statement will follow these criteria:

   • Concise and easily understood.
   • Begins with an action verb.
   • Describes an achievement that is measurable, observable, or demonstrable.
   • Assigned to one person for implementation, provided that person has the resources (human, financial) to accomplish the result.
   • Makes a significant contribution to the strategy (and therefore the mission and goals).
   • Contribution to the mission is worth the time, effort, and resources necessary to implement the result.

4. Develop a cost benefit analysis to forecast expenditures for implementation and sustainability.

5. Identify 5-10 high-level action steps/tasks required to complete the action plan. The PDSA cycle can guide writing the steps/tasks for the action plan.

   • **Plan** - What will it take to implement the program/initiative? What steps will initiate the project?
   • **Do** – How will we implement the program/initiative?
   • **Study** - What evidence will we collect to monitor implementation of the program/initiative? Who will collect the evidence? How will we report progress and when?
   • **Act** - What steps will we take to maintain and sustain the initiative? How will we evaluate effectiveness?

Monitoring and Reporting Progress

The monitoring and reporting process for each action plan provides an effective system of accountability and a means to monitor performance. Conduct periodic reviews with the team to check the progress of the action plan. Frequency may depend on the nature and complexity of the action plan.

Figure 5.4 Process to write action plans.

"There's a huge difference between the opportunity to '*have your say*' and the opportunity to be *heard*."[7] To be heard requires a culture where individuals feel safe and have a level of trust that they can speak up. A red flag flies up when I sit it on a meeting where no one speaks. Experience tells me that probably everyone in the room has something to say. They will say it in the parking lot, not in this meeting. Jim Collins offers leaders four practices that contribute to a climate where people can speak, and speak the truth[8]:

1. Lead with questions, not answers
2. Engage in dialogue and debate, not coercion
3. Conduct autopsies without blame
4. Build "red flag" mechanisms

It will take this type of climate and leader who can and will ask the hard questions to understand if the strategy we developed works and leads us to our goal. This leader and the leadership team must accept responsibility and willingly bare the brutal facts—good or bad. Without this level of trust and commitment, the process for reviewing your strategy in Figure 5.5 may only represent boxes to check rather than an opportunity to continuously improve and transform the organization to excel in what they do best.

Reviewing strategy at appropriate intervals maintains focus and the ability to adjust actions to ensure achievement of goals and overall strategy. How often should you review progress? That depends on the nature and urgency of the goals and the metrics you have chosen to monitor performance. At a minimum, conduct quarterly reviews by the senior leadership and Board. At the department level, reviews may occur more frequently to ensure adequate progress of your action plans. We will learn more about how to measure progress in Chapter eight.

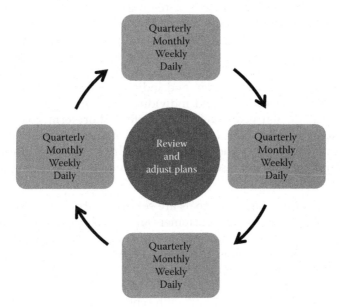

Figure 5.5 Process to review strategy.

What if?

The status quo is persistent and resistant.[9]

Seth Godin

What would happen if you stripped the mystery from developing and executing strategy? What could you achieve if you wrote an action plan that could guide the work? How much more could you accomplish if everyone could speak up and use meaningful data to review progress?

What if we committed to focus on our mission, align all the arrows, and integrate the work of our system?

You would assuredly transform your work and create a culture where excellence prevails. The leaders I interviewed (see Chapter eleven) work every day on doing just this and making remarkable progress. Are they all where they want or know they can be? No, remember, this is a process, not an event. But every one of them has a strategy, goals, action plans, and key measures of performance. They each lead their organizations, divisions, departments, or schools focused on their mission and with a steady eye on the goals that bring value to their customers and the people within who do the work to create excellence.

Insights for the journey

Anna Prow: For a number of reasons, many nonprofits may never have learned how to articulate what they're seeking to achieve, what problem(s) they are trying to solve, or what success looks like, so I start by sensitizing staff to think about our work in those ways. I work to develop this awareness at all levels—organizational, team, and individual. Then I work with staff to develop team and individual goals in support of the organizational goals we've articulated. Short-term goals are often easier for staff to establish and pursue. For longer term goals, I try to use a budget process as a driver so team members can imagine what resources it will take to achieve their goals farther out.[10]

Rick Rozelle: Long-term goals are set by the strategic planning process which engages customers and stakeholders. They start by getting clear on the mission of the organization (why it exists) and vision (how things will be better in the future). The vision is a broad statement of where the organization is going, and the goals are developed such that they refine and further define the vision. The goals represent the thoughts of a broad constituency. Short-term goals are established by identifying the types of objectives that further define and refine the goals. Objectives are more near-term, time-based and measurable. These are set by the leadership team that will be responsible for carrying out the objectives.[11]

JoAnn Sternke: I believe leadership is the work of the heart, head, and hands. A hundred percent of the people who go into education have the work of the heart pretty darn clarified. We know our purpose. It's our job to make sure that, as leaders, we focus everyone on the mission—work of the heart. You take the work of the heart and put it in the work of the head, which is planning. It's putting that plan in place that takes the work of the heart and makes it something understandable and put forth in a way that you can then achieve the work of the heart. Work of the heart, work of the head, but then it's executed with the work of the hands. What are the processes, the action plans? It can't stay in the heart. I began as a leader that had all heart. I really didn't understand the importance of putting it into a framework or a plan to give it intellectual cadence and clarity. But it also can't live there. That's when strategic plans live on a shelf. It can't stay in the work of the head. You need to then take it to the work of the hands. The work of the hands is what we do every day. We take that work of the heart that we put into an operational plan, and we execute it with work of the hands. That includes good processes and action plans. That's what makes the work real when you take the work of the heart, make it work of the head, and bring joy to the work of the hands by putting it in living motion. That to me is the best part of what we do as leaders.[12]

Notes

1. J. P. Kotter. BrainyQuote.com, Xplore Inc., 2017. https://www.brainyquote. com/quotes/quotes/j/johnpkott166628.html, accessed February 1, 2017.
2. L. Bossidy and R. Charan, *Execution: The Discipline of Getting Things Done* (New York: Crown Business, 2002), 8.
3. J. C. Collins and J. I. Porras, *Built to Last: Successful Habits of Visionary Companies* (New York: Harper Business, 1994), 87.
4. P. Bregman, "Execution Is a People Problem, Not a Strategy Problem," *HBR. org*, January 4, 2017, https://hbr.org/2017/01/execution-is-a-people-problem-not-a-strategy-problem.
5. L. Bossidy and R. Charan, *Execution*, 21.
6. L. Bossidy and R. Charan, *Execution*, 23.
7. J. J. Collins, *Good to Great: Why Some Companies Make the Leap...and Others Don't* (New York: HarperCollins Publishers, Inc., 2001), 74.

8. J. J. Collins, *Good to Great*, 74–80.
9. S. Godin, *Tribes: We Need You to Lead Us* (New York: Penguin Books, 2008), 91.
10. A. Prow, interview by author, February 15, 2017.
11. R. Rozelle, interview by author, February 22, 2017.
12. J. Sternke, interview by author, February 22, 2017.

section two

Align the organization for excellence

> Ultimately, a high level of organizational alignment is essential for achieving increasingly better business performance results now and in the future![1]
>
> **Torben Rick**

You have a mission. You have goals and a strategy. You know your customer and what they need from your products and services. But *who* do you need to realize success? What skills, expertise, and attitudes must people possess to achieve the mission?

The interaction between the system design factors and human performance factors—our people—determine 80%–95% of the results in the organization. What policies will drive excellence? What processes will lead to the outcomes we hope to attain?

Aligning for excellence requires purposeful and intentional decisions regarding the qualifications of our people, our customers' expectations for quality service and products, the identification and improvement of critical processes, and measures to know if those processes work. Without alignment to the aim, we remain reactionary. We work hard, but our success may come more by accident than our intent. A culture of excellence aligns people, purpose, and processes to be and become our best—individually and collectively.

Notes

1. T. Rick, "Organizational Alignment Is the Glue for Achieving Better Performance," *Meliorate*, October 10, 2014, https://www.torbenrick.eu/blog/performance-management/organizational-alignment-is-the-glue/.

chapter six

Value people and their contributions

> Maybe the biggest thing we've forgotten in organizations is that human beings work for us.
> The most sophisticated strategy is worthless if humans can't embrace it or be engaged with it.[1]

Jim Haudan

How do you value people? The story that follows really happened. Everyone survived unscathed, but the lesson learned remains. How we treat others does matter. Our expectations should always remain clear. We need committed and dedicated people to transform our organizations. Without them, we will never get there.

The day began chaotic, and all the planning for the training session seemed to disintegrate piece by piece. Materials failed to arrive on time. The tech team struggled with projectors and computers. Her partner, Liz, had to take over another meeting when a colleague fell ill. Since they had designed and prepared the training together, Beth had no problem facilitating the session alone. Liz went on to the other meeting, and Beth went to inform the director.

Beth never hesitated to make decisions. She weighed the consequences of cancelling with a room full of participants or stepping in to lead the training. Beth walked into the director's office and explained the situation and the change in plans. She never expected what happened next. The director began shouting, "You did what? Making decisions is not your job!"

Beth, never one to back down, calmly responded, "I considered the impact on our customers who sat waiting for both events to begin on time and professionally. You hired me for my expertise and experience, and the ability to make decisions."

By the end of the day, the training and the meeting met the needs of those who came to learn and meet with peers to discuss new regulations. The director never mentioned the scene again. The sharp words would remain with Beth. She lost all respect for a leader who did not value or respect her enough to explain why the decision might have caused a problem—or apologize for the outburst.

People deserve respect. We thrive on respect. Taking that away erodes trust and can instill fear and uncertainty in the workplace. "Simply giving employees a sense of agency—a feeling that they are in control, that they have genuine decision-making authority—can radically increase how much energy and focus they bring to their jobs."[2]

Transforming our organizations depends on the expertise, commitment, and trust of the people who do the work. Jim Collins'[3] research found that great leaders recognize three truths:

1. If you begin with "who" rather than "what," you can more easily adapt to a changing world
2. If you have the right people, the problem of how to motivate and manage people largely goes away
3. If you have the wrong people, it does not matter whether you discover the right direction, you still will not have a great company (school, government, nonprofit)

We must never forget, "Great vision without great people is irrelevant."[4]

When we demonstrate that we value people and their contributions to success and excellence, we have little difficulty answering these questions:

- How do you explain the mission and help people understand how their work contributes to it?
- How do you create trust and eliminate fear in the workplace?
- How do you listen to employees without passing judgment?
- How do you build a work environment that supports and rewards cooperation?
- How do you model and encourage personal and organizational learning?
- How do you coach others for high performance and excellence?
- How do you know if people need help in improving or developing new skills?
- How do people know you value them?

What is engagement?

> When people are financially invested, they want a return. When people are emotionally invested, they want to contribute.[5]
>
> **Simon Sinek**

Motivation

Leaders set the stage for engagement. They clear the obstacles that impede our best work. When "certain conditions are met and people inside an organization feel safe among each other, they will work together to achieve things none of them could have ever achieved alone."[6] If we create an environment of trust, and we have the right people, we clear the path for intrinsic motivation that leads to a commitment to the work of the organization—engagement.

Sometimes, we believe that money drives commitment. If that is true, how do you explain well-paid individuals who remain unengaged even to the point of undermining leaders and their colleagues? When we focus solely on extrinsic rewards, we miss opportunities to show authentic appreciation. This does not mean that extrinsic motivators cannot contribute to employee satisfaction or engagement when used judiciously. But an overuse of sticks and carrots creates, at best, compliance. I can comply and never engage.

You can spot motivated people. They do whatever it takes to get the job done. Problem-solving energizes them. They ask questions and seek clarity. No one needs to prod motivated people. They take the initiative and rarely wait to be told what to do. They smile and seem to genuinely enjoy their work and their colleagues.

Intrinsic motivation, the drive that comes from within, yields commitment. Engagement requires commitment. "We must preserve the power of intrinsic motivation, dignity, cooperation, curiosity, joy in learning, that people are born with."[7]

Engagement

What is *engagement*? What does is it look like? What would you expect to see with engaged employees? Descriptors might include trust, satisfaction, empowerment, teamwork, or innovation. How does engagement differ from satisfaction? People may have a high level of job satisfaction based on their benefits or proximity to home, but engagement requires more than satisfaction.

Employees generally fall into one of three types:[8]

1. Engaged—committed to the goals of the organization, have a passion for the work they do, open to creative and innovative thinking to solve problems
2. Not engaged—just going through the motions "it's just a job," putting in time but without energy or passion
3. Actively disengaged—negative, may undermine goals of the organization, not a team player, not just unhappy at work but often busy acting out their discontent

You cannot decree engagement. Engagement emerges within a culture where we feel safe, know that we have value, and believe in the purpose of our work. Engagement cannot exist when we feel threatened or fear retribution for our actions. In his 1954 book, *Motivation and Personality*[9], Abraham Maslow identified a hierarchy of needs. He proposed that the most basic of needs must be met and mastered before moving to the next level. Figure 6.1 expands the types of engagement and compares it with Maslow's hierarchy.

When we look at engagement this way, we can begin to see the importance of meeting personal needs for safety/survival, security, belonging, importance, and self-actualization. People want to be a part of something important. They need clarity of purpose. All of us want to belong and know that we have value. We want to know that our work has meaning and leads us to excel and meet challenges. We want to make a difference, for each other and the customers we serve.

Employees want to know and understand the "why" of their work. They want their leaders to connect the dots. Supervisors, managers, frontline staff rarely, if ever, see the big picture. Their vantage point does not allow them a glimpse into all the factors that might influence decisions. Creating clarity opens communication and trust between leaders and employees. Answering just a few simple *why*[10] questions can mean the difference between opposition and commitment to the mission.

1. Why is it important?
2. Why me (and not someone else)?
3. Why now?

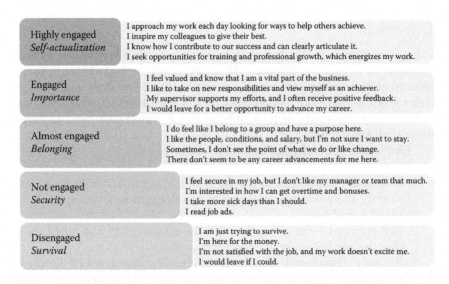

Figure 6.1 Employee engagement and Maslow's hierarchy of needs.

4. Why do it this way?
5. Why would I want to do this?

Continuous improvement inherently involves change. Think of the difference in how change efforts that respond to these questions fill the need for safety, security, belonging, and importance. We can accomplish so much more by simply answering *why* and connecting the dots so that everyone sees what you see as the leader. We should not forget that our success depends on people "if employees truly are a company's best asset, then leaders and managers should make caring for them a priority."[11]

What is a high-performing team?

> The way a team plays as a whole determines its success. You may have the greatest bunch of individual stars in the world, but if they don't play together, the club won't be worth a dime.[12]

> **Babe Ruth**

Humans are social. We come into the world hardwired to survive, find our tribe, and discover where we belong. The notion that you can work better alone may apply some of the time, but not all the time. "Two heads are better than one," may sound trite, but we create and learn more collectively than we ever could alone.

I like reflecting and spending time with my own thoughts. Something extraordinary occurs when I share those thoughts, and we mull over lofty and not so lofty ideas together. An element of synergy that I never find individually suddenly explodes when I sit with colleagues to discuss, digest, and even disagree.

High-performing teams experience a level of exceptional synergy and productivity. Many, perhaps most, teams drown in complacency, conflict, and competitiveness. What makes the difference? Why would I relish every moment working with one group of people and dread the company of another? This topic goes well beyond the scope of this book, but we can explore a few characteristics exhibited by great teams.

I hope you have had the opportunity to work on a team where each person brought their best to the table every day. Where no one withheld their knowledge, experience, and certainly not their opinion. Some of us have the good fortune to serve on many highly collegial and interdependent teams. As of this moment, three superior and dynamic teams come to mind through my professional career.

One team included a diverse group charged with creating and delivering training at various locations for different functions and levels of staff. The first day we met, my immediate thought was, "This will never work!" Twelve women of different ages, expertise, experience, and personalities. We ranged from outspoken to quiet, but we all shared a common vision. Our leadership gave us a simple goal—to build our internal capacity to offer professional learning.

You can call it luck, but the accomplishments of this team far exceeded sheer luck. We were set up for success from the beginning. The members completed a comprehensive application process that included an essay on why we wanted to take this step in our careers and a video demonstrating our teaching ability. I guess you could argue that this alone narrowed the field by selecting highly motivated and skilled individuals. That argument would hold if I had not seen teams also carefully chosen for expertise that never rose to that level of collaboration.

We completed several trainings on adult learners, presentation skills, and shadowed other experts in the material that we would teach. We experienced a Ropes Course[13] that helped us individually and as a team by exploring leadership and communication concepts, problem-solving, and coaching. The high ropes challenge tested our level of risk taking and trust in the facilitators and our colleagues while dangling precariously in the air. We spent that day learning about ourselves and each other.

Every team does not need or has the resources for a full day of activities like the Ropes Course. The course did not guarantee that our team would work well together. The time spent added one more dimension to building our success. In a relatively short period of time, we became more than any one of us could have become individually. The work we created and the training we delivered far surpassed anything we might have designed and delivered individually.

Could any of us have taken the material and taught successfully? Absolutely. By the end of three years, we had touched nearly all the 500 employees, and even won over a few who entered with folded arms. We did more than teach. We modeled. We mentored. We coached. We supported learning—the participants' and ours. While we all gained something valuable, our customers came out the real winners. Everyone had new skills to tackle the challenges we faced.

Common elements shape high-performing teams made up of quite ordinary people. We often do not have the luxury to pick our teams, nor do we know why we received the call to serve on *this* team. When I recall the good and the not so good teams in my past, they either excelled or failed miserably at any combination of a few key characteristics. High performance teams:

1. Align around a shared vision and values
2. Feel safe and nurture an environment of trust
3. Exhibit integrity—doing the right things for the right reasons
4. Respect one another

   5. Remain learners and encourage new ideas and thinking
   6. Give and receive beneficial feedback
   7. Communicate well and often—open discourse and dialogue
   8. Collaborate, complement, and challenge each other
   9. Value the strengths each member brings to the team
  10. Focus on attaining goals

   No perfect team exists, but we can work toward creating teams based on those 10 characteristics. Leaders play a vital role in establishing the climate and culture where teams perform well and get results. Remember, we have a social nature and gravitate toward groups. Unfortunately, no one can make people, who did or did not choose their place on the team, cooperate and work well together. Leaders have one critical responsibility in building an environment for teams to excel—"remove the system processes (barriers) that derail team-building… *remove organizational practices that damage teamwork.*"[14]
   What do you look for to know if you have a problem within a team? When teams flounder, it usually signals a lack of alignment. Alignment in the sense that members may not all agree with one another. Agreement can often stifle team thinking. If everyone in the room agrees, we might question the level of thinking of the group. Creative and productive teams challenge thinking, and disagreement often results in the best product. How many of these seven signs of a lack of alignment did you see in your last team meeting?[15]

   • People remain silent and do not voice their opinion when you call for a decision
   • You keep being surprised by the actions people take because they are inconsistent with the agreed-upon direction
   • You do not see tangible progress on an issue when by all rights you should be moving forward throughout the organization
   • In meetings, people keep bringing up issues that you thought were resolved
   • People complain, make excuses, and blame others for lack of progress
   • You observe a lack of ownership and enthusiasm for implementing a course of action that has been set
   • People state that they disagree with a decision or direction that has already been taken

   Leadership teams play a significant role in the ability of an organization to execute strategy and support a culture of excellence. One definition of a leadership team describes it as "a small group of people who are collectively responsible for achieving a common objective for their organization."[16] Corporations, departments, churches, or small nonprofits thrive when the leaders at the top form a cohesive team.

If leaders do not align themselves to the mission and goals they have established, how can we possibly expect our employees to believe they are important? Leadership teams undermine their work and the entire organization if they walk out and express doubt or criticize the decisions of the team. I have seen this happen and more than anything else, we create confusion and distrust. We want our leaders to lead. If they all go in different directions, who do we follow?

How do you create an environment that values people and their contributions?

> The most important thing in communication is hearing what isn't said.[17]

> **Peter Drucker**

The processes that follow may appear obvious and expected of any worthwhile company. Most organizations hire, train, and evaluate employees. The difference lies in the alignment of those critical processes to the goals and mission. This subtle, but vital, difference can contribute significantly to transformation. How intentional are we in bringing on the right people, providing learning opportunities, and understanding how individual performance supports excellence?

Hire staff

How do you determine who should join your organization? Do you look just at credentials and experience, or do you probe for underlying perceptions and values? Jim Collins[18] talks about getting the right people on the bus, and equally important, getting them in the right seats on the bus. Recruit the right people, but also keep them because, "if we don't re-recruit and develop them (92% who meet expectations), some will leave. Never take a person for granted."[19]

Do you have a systematic approach to hiring? When we do not take the time to do this well, we may or may not put the right people in their seats. We do a disservice to our company and to the person we brought on the bus. People will not change their personalities or values just because we say they should. Marriages end for just this reason. We become enamored and overlook faults and values that later raise their ugly head. We cannot turn back without a lengthy or nasty divorce. Hire for the long term and build relationships. Divorce in marriage and business does not look pretty, causes pain, and results in loss of resources.

Take a deliberate look at your current practices and evaluate the extent to which you currently follow a consistent process for hiring people

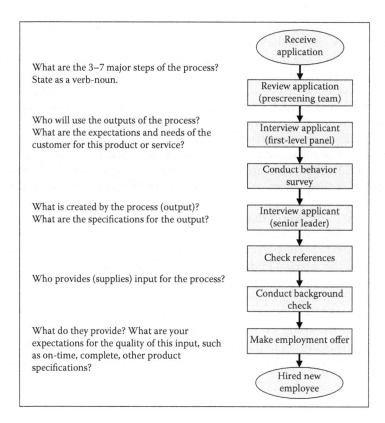

What are the 3–7 major steps of the process?
State as a verb-noun.

Who will use the outputs of the process?
What are the expectations and needs of the
customer for this product or service?

What is created by the process (output)?
What are the specifications for the output?

Who provides (supplies) input for the process?

What do they provide? What are your
expectations for the quality of this input, such
as on-time, complete, other product
specifications?

Figure 6.2 Process to hire staff.

who bring expertise but also align to your mission and values. Figure 6.2
shows the process at a high level. Depending on the size and nature of
your business, each step may include one or several subprocesses. While
the steps and sequence of the process are important, the more significant
differentiator of excellence lies in knowing you have consistency in what
you do. You also know where to start looking if things do not go well.

Develop the workforce

Developing the potential in people gives them more than just hours in
training. Building the capacity of our people builds the capacity of the
organization. Creating the pathway to excellence, and then executing
specific actions along the path needs "the right people, individually and
collectively who focus on the right details at the right time."[20] They need
the skills for decision-making, the ability to think incisively, and have the
capacity for realistic and candid dialogue. To effectively reach goals and
execute strategy, everyone must have these skills, not just a privileged few.

Our success depends on the ability of the CEO and frontline workers to think, speak, and make sound decisions.

Even when we hire the best, we must continue to help them become their next level best. We build learning organizations through individuals and leaders who continue to learn every day. I have listened to leaders who do not think they need to continue to learn. After all, they have acquired years of experience.

That logic has two serious flaws. We cannot always learn from experience since the consequences of that experience may be far removed from the event. Reflection on an experience often deepens our insight into the potential consequences or outcome of our experiences. The second problem—the world changes—we may need new tools, new understanding to operate in the world that will exist tomorrow or a year from now. The process in Figure 6.3 considers the needs of each employee and uses that information to create personalized opportunities that benefit the employee and build the internal knowledge bank of the organization.

Learning and building capacity tightly aligns to the next process, evaluating individual performance. Amassing hours of training does not necessarily equate to increased skill or knowledge. "Learning is not attending, listening, or reading. Nor is it merely gaining knowledge. Learning is really about translating *knowing* what to do into *doing* what we know. It's about changing. If we have not changed we have not learned."[21]

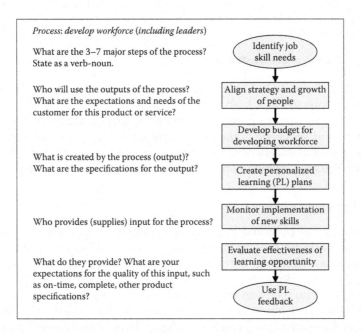

Figure 6.3 Process to develop the workforce.

Manage workforce performance

John waited outside the office trying to anticipate how his supervisor, Carol, would evaluate his performance. No doubt this would be the same as every other year. He would walk out the door at the end of the meeting knowing little about what she thought he did well or how he could improve. How much can you learn about yourself from a five-point Likert scale of irrelevant statements?

Even if Carol did not ask, John had reflected on his strengths and accomplishments. He brought the list to the meeting prepared to share his insights. As he walked in the door, Carol seemed distracted. "John, great to see you. Have a seat. This shouldn't take too long."

"Well, so much for having a discussion," John mused as he quickly put his notes back in his folder.

Carol went down the list of expected behaviors, and as usual, John received "outstanding" for most of them, except for two areas. "John, I think you should work on how you communicate across departments, so I have scored that item lower."

John thought about the statement a moment. He had worked collaboratively across several departments in the past year, which resulted in cost savings and greater integration of their work. "Can you explain what I failed to do in this area or give me an example of what you are looking for?"

"No, nothing specific. I just see that as an area of need across the department." Carol signed the evaluation form and handed it to John. "I have to run to another meeting. Just keep up the good work. Sign two copies. One for you and one for the file."

John thought to himself as he walked back to his office, "Maybe I should start looking at other opportunities. How much more can I do, and does it even matter?"

This scene repeats itself day after day, year after year, and we wonder why we lose employees or they seem unmotivated and disengaged. "The best performance management programs are designed to stimulate the right kinds of conversations around the right topics. That's all."[22] The interaction between John and his supervisor would hardly qualify as stimulating conversation on any topic.

We do not need an elaborate system, but we do need a human system that values the work we do each day. If we build and implement clear direction and give people the freedom to do their best work within a set of clearly communicated boundaries, they will do their best work. When we hire the right people, it frees leaders to manage the system rather than people.

All of us want to know how we are doing in the quality of our work and level of productivity. We want support and nonthreatening ideas on how to improve. We want to be seen. Coaching, mentoring, and

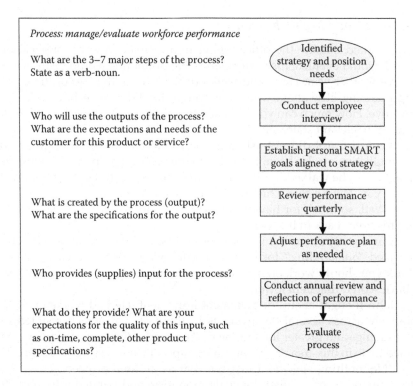

Figure 6.4 Process to manage workforce performance.

meaningful conversations ensure that we take the time to see our employees and understand how to clear the path for their professional growth.

Consider how you can adopt or adapt the process in Figure 6.4 to begin to move from a perfunctory annual performance evaluation to more frequent or even ongoing opportunities for employees to demonstrate their commitment to the mission, their part in accomplishing organizational goals, and what they want to become. They will surprise us in their ability to accurately and honestly assess their own performance. Some will try to fool us—but only because they do not yet know how to reflect on their work or feel safe to speak honestly about themselves.

What if?

Valuing people begins with respect and building positive relationships. We rarely trust those we do not know or whose motives appear self-promoting. What if we took the time to design hiring practices aligned to our mission and values? What if we encouraged learning and allocated the resources to develop and re-recruit our best and those whose potential just needs a push. What if we engaged in meaningful dialogue and reflected on our performance? What if we would "constantly bridge the gap between people and possibilities by knowing—and practicing—what it takes to tap into the latent, unused potential that's just waiting to be awakened and engaged."[23]

Insights for the journey

Trent Beach: I listen to staff in regular team meetings. The structure of my meetings is to speak about 10% and listen about 90%. This is also how I do visits with staff during the workday. These are prescheduled on my calendar so that even as appointment requests come in, these times to "get to the shop floor" and listen to my team members are protected.[24]

Debra L. Kosarek: I listen to staff in terms of not only what they say, but how they say it; and perhaps, more importantly what they do not say. I have found it is true that leaders that stop listening are eventually surrounded by people that have nothing to say. By periodically bringing up the goals one on one, as well as with the group, they know I consider the goals important. They know they are "on my radar," and not just something completed and put on the shelf.[25]

Anna Prow: In large part, I listen to staff by getting to know them personally and caring about their lives, their work, and their contributions. As for their support of my goals, I find they are more committed to my goals if they truly believe that the accomplishment of my own goals involves value for them— whether that be facilitating their success or improving conditions for them and the team overall.[26]

JoAnn Sternke: Similar to how you listen to community. Engage staff in a very strategic way. It's all about respect and listening. Treat them as partners. I open doors, listen, learn, you put things in motion, and remember that it is our employees that live off the promises that I make every day. They are our most valued treasure. Our community knows it. We know it. We make sure our employees are respected and listened to because they are the ones who carry out our mission most poignantly. I make sure that my entire team understands what it means to listen to staff. It's not just nodding but also letting them know you reflected and you value their opinion because you can't do everything. That creates a respectful working climate. Tend to culture and remember that it is not about you. It is about those you serve.[27]

Notes

1. J. Haudan, *The Art of Engagement: Bridging the Gap between People and Possibilities* (New York, McGraw Hill, 2008), 242.
2. C. Duhigg, *The Power of Habit: Why We Do What We Do in Life and Business* (New York: Random House Trade Paperbacks, 2012), 151.
3. J. J. Collins, *Good to Great: Why Some Companies Make the Leap... and Others Don't* (New York: Harper Collins Publishers, Inc., 2001), 42.
4. J. J. Collins, *Good to Great*, 42.
5. S. Sinek, Twitter post, May 22, 2012 (3:27 p.m.), accessed January 6, 2017, https://twitter.com/simonsinek.
6. S. Sinek, *Leaders Eat Last: Why Some Teams Pull Together and Others Don't* (New York: Penguin Group, 2014), 15.
7. W. E. Deming, *The New Economics for Industry, Government, and Education*, 2nd ed. (Cambridge: Massachusetts Institute of Technology Center for Advanced Educational Services, 1994), 121.
8. R. Reilly, "Five Ways to Improve Employee Engagement Now," *Gallup Business Journal*, January 7, 2014, accessed April 13, 2014, http:/businessjournal.gallup.com.
9. A. H. Maslow, *Motivation and Personality*, 1st ed. (New York: Harper & Brothers, 1954).
10. R. Connors and T. Smith, *How Did That Happen: Holding People Accountable for Results the Positive, Principled Way* (New York: Penguin Group, 2009), 159.
11. R. Reilly, "Five Ways to Improve Employee Engagement Now."
12. B. Ruth. BrainyQuote.com, Xplore Inc, 2017. https://www.brainyquote.com/quotes/quotes/b/baberuth125974.html, accessed September 18, 2016.
13. "Ropes Course," *Wikipedia*, last modified January 8, 2017, https://en.wikipedia.org/wiki/Ropes_course.
14. L. Jenkins, L. O. Roettger, and C. Roettger, *Boot Camp for Leaders in K-12 Education: Continuous Improvement* (Milwaukee: ASQ Quality Press, 2006), 11.
15. R. Connors and T. Smith, *How Did That Happen*, 172.
16. P. Lencioni, *The Advantage: Why Organizational Health Trumps Everything Else in Business* (San Francisco: Jossey Bass, 2012), 21.
17. P. Drucker. BrainyQuote.com, Xplore Inc, 2017. https://www.brainyquote.com/quotes/quotes/p/peterdruck142500.html, accessed March 15, 2017.
18. J. J. Collins, *Good to Great*, 42.
19. Q. Studer and J. Pilcher, *Maximize Performance: Creating a Culture for Educational Excellence* (Pensacola: Fire Starter Publishing, 2015), 43.
20. L. Bossidy and R. Charan, *Execution: The Discipline of Getting Things Done* (New York: Crown Business, 2002), 34
21. J. G. Miller, *QBQ! The Question Behind the Question: Practicing Personal Accountability at Work and in Life*, (New York: G. P. Putnam & Sons, 2004), 110.
22. P. Lencioni, *The Advantage*, 164.
23. C. Duhigg, *The Power of Habit*, 242.
24. T. Beach, interview by author, February 17, 2017.
25. D. L. Kosarek, interview by author, March 4, 2017.
26. A. Prow, interview by author, February 15, 2017.
27. J. Sternke, EdD, interview by author, February 22, 2017.

chapter seven

Identify key work processes

An organization is only as good as its processes.[1]

Geary A. Rummler and Alan P. Brache

Every one of us interacts with and works within processes each day. Processes develop by design or by default. Many times, our processes work because of sheer luck—without any clear intent other than to get from point A to point B. We often do not notice a process until things go wrong or work flow does not function smoothly.

Stand in line at the post office to buy stamps and just watch. You might be fortunate enough to walk in and go right to the counter. I rarely have that opportunity. Instead, I find myself about the sixth person in line behind the person with eight large packages. I have plenty of time to assess the process—mine, as the customer, and the postal worker at the counter.

Some days, the process works smoothly. Even with a long line, transactions occur quickly. I do not pay much attention to the process, I just keep moving up in the line. But some days, I have more than enough time to assess, analyze, and consider improvements to the process. During the holidays, one of the busiest times for the post office, we all stood in the same line. Whether you had packages or just needed a stamp, you all stood in the same line. Where is the express lane, and why would you not create a designated line for packages?

A simple change in the process would have allowed over half the people in line to buy stamps and leave within five minutes. We encounter processes everywhere. We get dressed in the morning—a process. We load and unload the dishwasher—a process. We hire new employees—a process.

Many processes begin with the output of another process, and your process often feeds into the process of another department or function. Nothing happens in a vacuum. In a system, the effectiveness of one process affects the quality of another process or even several processes. Transforming an organization and striving for excellence ultimately lies in the ability of our processes to create quality products and services and answering a few questions, such as

- How do you determine your organization's critical processes?
- How do you know that the processes in place support the aim of the system?

- How do you know they align to your strategy?
- How do you know they work?
- How do process outputs move to the next process?
- How do we handoff our work to the next person, the next department, the next machine?
- If the process broke down what would be the impact to the company?
- Do leaders embrace the role of process owner?
- How do you manage processes?

What is a process?

> If you can't describe what you are doing as a process,
> You don't know what you are doing.[2]

> **W. Edwards Deming**

Simply stated: A *process* is a series of tasks or activities that take an input (those things needed to do a job), modifies the input (work takes place and/or value is added), and produces an output (service or product). Identifying every organizational or individual process may be noble, but is not the goal or particularly practical. However, those processes that critically affect success will be our primary focus. A *critical process* is defined as a process that is *essential* to the accomplishment of organizational goals and objectives.

Consider processes in the context of a system as shown in Figure 7.1. The system has an aim and usually contains subsystems or functions, such as finance or human resources. We can look at processes from three

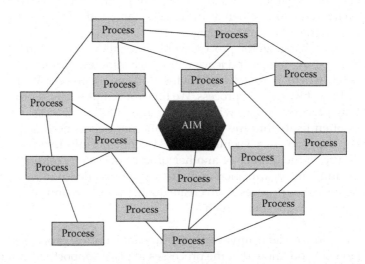

Figure 7.1 Interrelationship of processes to the aim of the system.

levels. The design and management of systems and their subsystems is generally strategic in nature. Processes occurring at the department or function level are often cross-functional and *tactical*. Every process consists of delineated *tasks* or projects within the process. These *operational tasks* or steps form the daily operations of departments or functions.

Systems, with an aim, consist of many interrelated processes. Processes do not occur in a silo and are cross-functional either across the organization or even across a subunit. Processes always occur horizontally. When we forget that simple fact, we work with blinders on. We do not understand that our process depends on input, which is an output of another process, or that someone else depends on the quality of our output as the input for their process.

How do you do something better? Stop for a moment and consider the last time you wanted to do something better than you had before. What did you do? If you hope to swim competitively, you may start by taking lessons and practicing different strokes. Both activities may yield some improvement in your swimming, but serious competitive swimmers go one step further. To improve their form, they must know what goes into each move, practice doing it, and then apply it to get to the ultimate goal—winning the race.

Whether it is swimming, cooking, paying the bills, hiring employees, cleaning the building, or planning a lesson, improving performance requires identifying and understanding the process of how we accomplish the work. A process is a planned and repetitive sequence of steps and activities for the delivery of a service or product. Every process has inputs, something or someone supplies the process, and outputs, something or someone receives or benefits from the process. Processes can be simple or exceedingly complex, and some are critical processes or functions that make up essential activities required to achieve our mission or goals.

Processes, whether we overtly define them or not, play an important role in how we do our work. Their design establishes the boundaries, which control the range of potential outcomes. If you do not like the current outcomes, look first at the processes responsible for that outcome. Two questions regarding processes should remain at the forefront:

- Does what we are doing add value?
- How can we improve results if we ignore the systems/processes by which we achieve those results?

We must always keep in mind that processes live in systems. Within that system, other subsystems and processes compete for resources. The different priorities within the system affect processes. Are the arrows, processes, aligned and integrated, or do they bump against each other creating cross-purposes and conflict?

When we try to fix one problem, we can potentially create problems elsewhere in the system. Well, you might say, "It works for us to do it this way." Without any malicious intent, we unwittingly create havoc somewhere else. I cannot overstate this fact: All processes occur horizontally across the system.

High levels of performance in swimming require more than practicing a single swim stroke. Although a swimmer may compete as an individual, they have a team that includes a coach, other swimmers, and supporters who collaborate to prepare for outstanding performance. Similarly, process improvement takes a team. Since processes are largely cross-functional and defined by the system, choosing the membership of a process improvement team becomes critical to the success and implementation of improvements to our work.

Why is becoming process-focused important?

> All work in all organizations is powered by processes.[3]

Rod Napier and Rick McDaniel

Do you have a process-focused organization? I have a quick assessment that takes very little time or background knowledge. Ask this question of yourself and others, "When something goes wrong, do we look for problems in how we do the work, a process, or do we look for someone to blame?" Do we look first at who did this and what training they need to fix their inadequacy? Process-focused organizations ask, "What went wrong in this process? What step did we miss? Do we have the right input?"

Recall our discussion in Chapter two regarding systems (Figure 2.8). We can attribute over 80% of the results to the system design and the interaction of that design with people. People generally contribute 20% or less, individually, to the outcomes of the organization. Deming noted that over 90% of work problems are not a function of individual people but the processes in place. If you see a problem or inefficient result, look first at the processes in place. Dynamic and complex systems contain a myriad of processes. Since processes involve suppliers and customers of the process, continuous process improvement must involve all levels and functions of the organization. No process ever sits in isolation.

Have you ever been in a work group that was wrangling with revising some form that no longer supported the work to be done? How many hours did the team argue on what should and should not be on the form? How often did the team end with making only minor changes, or worse, added more to meet everyone's needs?

I am often asked to help dig a team out of the mire of their paper-filled world. Seeking to streamline their process, the expected culprit is usually "the form."

This work typically begins with bringing some form that has been the source of much contention and ineffectiveness. The problem is that the form is often only the symptom of a faulty process. We can revise the form all day and still not address the primary issue—a broken or ill-defined work process.

Need an example? Most organizations have a form for approving new positions. Does your company have such a form? How many sign-offs do you require for the form? How long does it take to route the form through all the approval levels? Does the form get lost in a black hole of inefficiency? Does anyone really read the form, want the form, or use the form?

You would be surprised at the responses I have heard. Here are a few:

"We don't really use the form, but Ms. Needtoknow (who has no authority over hiring) insists on signing off on all new open positions."

"I don't use that form, I created another one because I needed more information."

"That form? I just call HR. I would never get any new positions if I had to go through that process!"

Any of that sound familiar? We have filled our day with busy work and then wonder why the system is so slow, bureaucratic, and filled with mistrust.

We start with the wrong issue—the form. The form may need improving, or we may not need it all, but how do we do the work? What is the work, and how do we know we designed the work to meet the needs of the customer or end user?

Many times, the form itself becomes the bottleneck. Required by "someone," the form arrives on your desk incomplete, inaccurate, or late. Now, you must stop your work and track down the missing information, make corrections, or email, call, or beg for the "form." Then the blaming begins.

Are you beginning to see the vicious cycle we create? The problem may or may not be a poorly designed form that does or does not support the process. The problem is not the person submitting the form or processing the form. The problem is not the thing or the person. The problem is the process.

When something does not work well or is not accomplishing the intended outcome, try these four questions:

1. What is it you are trying to do?
2. Who is it for, and what are their expectations?
3. How are you doing the work to reach the outcome?
4. How do you know that what you are doing is meeting the outcome?

Attributes of a process-oriented culture

- Departments are partners, not competitors.
- Departments, and ultimately, individuals) are measured on their contributions to process effectiveness and efficiency.
- Upstream, as well as end-of-line, performance is measured and tracked.
- Resources are allocated based on process needs (not results).
- Managers "manage the white space" rather than managing individuals.
- People understand their department's suppliers and customers and how their department fits into the "big picture."
- Individuals are encouraged to communicate directly with their peers in other departments.
- Cross-functional collaboration is rewarded.
- Problem solving focuses on finding and removing the root causes of disconnects in the system, not on addressing symptoms or fixing blame.
- Cross-functional teams routinely address major issues.
- Customer needs and concerns dominate decision-making.

Figure 7.2 Attributes of a process-oriented culture. Reprinted with permission from Bryan R. Cole, EdD.

Most of the time we have a process problem—not a people problem. You may tell me that I just do not know this person. If I did, I would clearly see the cause of the issue. Is it really a people problem? What processes do you have for developing accurate job descriptions, hiring, coaching, and developing skills? If the person consistently undermines and sabotages the work of the team or just won't work, do you have a process to respectfully remove them from the work place?

A process-focus allows teams to analyze and diagnose all types of problems without finger pointing and unfairly placing blame on people. Most organizational problems have their root cause in a process. A process mindset helps us manage our work more effectively and efficiently. Figure 7.2 describes the attributes of a process-oriented culture. How many do you find in your work place?

Why should we map processes?

> The map of a work process is a picture of how people do their work. Mapping is merely an enabler—a means to a more important end.[4]
>
> **Dianne Galloway**

What is the more important end? We want to capture the "how?" People bring creativity and innovation to their work every day. Those same people "work around" outdated, ill-defined, or unspecified processes. They make it work because that is their job. Even if it takes more time, more energy, and a great deal of daily frustration, people care about the quality of their work. They want to give their best. Unfortunately, the processes in place can create obstacles and barriers. Our best employees do the work anyway.

Process maps

A simple map or picture provides a view of "what is" in all its glory or ugliness. When we want to improve how we do our work, a map can show us how it "should be." You will find different definitions for a flowchart or process map. For purposes of this book, the terms are interchangeable. Both give us a picture of our "how."

Process maps show the separate steps of a process in sequential order and provide a visual of the process. They provide valuable information about how the process works. Knowing the steps facilitates the identification of where or how we can improve the process. A well-defined flowchart also serves as a communication tool for members of the team who are involved in the process. It gives the team a common understanding of their work.

Documenting a process and publishing that process for others to review develops a deeper understanding of how processes flow to and from each other. This becomes critical to building an integrated work system—a system, where work processes complement and support each other rather than compete for valuable resources. Through a map, everyone understands the project from beginning to end.

Process mapping steps

You do not need any sophisticated software or extensive quality background to create a flowchart of your process. I usually use square sticky notes, felt tip markers, and a large sheet of white paper. Even a wall or table top will serve the purpose. Computer software can make the final product pretty and give you an electronic version, but the basic hands-on materials work best for the initial stages of creating the process map.

The work can be messy, but fluid, and allows time to get everyone's input and perspective. There is no one right way to map a process. What is important is creating a map that helps those involved understand the process.

Basic flowchart symbols help illustrate what happens in the process. You just need four simple symbols: an oval, a rectangle, a diamond, and an arrow. The process map begins and ends with a terminator, the oval, which reflects the starting and ending points. I usually just draw this on the chart paper. This sets the boundaries for the process.

Place each step in the process in a rectangle, the square sticky notes. Use an arrow to show the flow of the process. Decision points based on a question appear in a diamond shape and require a "yes" or "no" or some other choice that demonstrates alternate paths for the process.

Other symbols can be indicated, if the process has a step that requires waiting for additional input or a delay of some kind. Sometimes, you may want to draw attention to a specific input or output within the flow of the process. Similarly, you can reference documents that may be a part of the process. Many processes feed into others, or we must complete a separate process prior to, or parallel with, your process.

You can show these on your map using the symbols in Figure 7.3. At the beginning, just focus on the four primary symbols. Everything else can add context and details that you may find important to documenting and communicating the process. Several key steps will guide the development of your process map.

1. Define the process (process objective and SIPOC)
   When you developed the process objective and completed the SIPOC, you identified the boundaries of the process. Where or when does the process start? Where or when does it end?

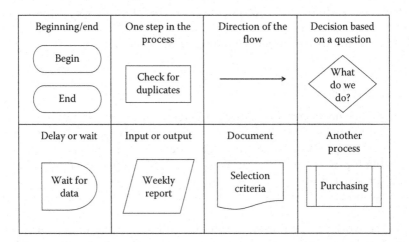

Figure 7.3 Basic flowchart symbols.

2. Brainstorm the activities/major steps that take place. Write each on a separate sticky note. Sequence is not as important at this point. Ask "what, where, how, and who" questions. Now is not the time to ask "why" questions. That will come later in your analysis.
3. Arrange the activities in proper sequence.
4. When you have identified all the activities or steps and everyone agrees on them, draw arrows to show the flow of the process.
5. Review the process map with others involved in the process such as suppliers, workers, or customers. See if they agree with how the process has been drawn.

Keep in mind that an "as-is" process map shows every step in the process just as it occurs. What can you expect from this activity? You realize everyone does the work differently. Key steps take place in another department. The process has become convoluted with multiple decision or approval points and constant repetition of steps. Laying out the "how" begins to illuminate the reasons for bottlenecks, rework, or poor quality. We begin to understand and identify opportunities to improve how we do our work.

Always take time to identify any other quality issues that may affect your process such as necessary resources, barriers, or other problems that may hinder the process. Allow time for in-depth dialogue about the steps in the process to ensure accuracy of the process map. And remember, the primary reason for a process map is to identify areas for change or improvement of the process. The following questions can help the team check the accuracy of the process flow:

• Is the process step the same for special circumstances?
• What happens if a substitute or new employee operates the process?
• Are there unexpected changes that might occur from the supplier or customer?
• Do some people do things differently? Is there variation in the process?

The answers to these questions emerge as the team begins brainstorming the steps. Process maps help reduce or eliminate inconsistencies or variation in the process.

Figure 7.4 shows a flowchart for how to process an order. Note that the process begins when the email with the purchase order arrives. The first step in the process is to put the information in a database. Before the next step, there is a decision point. Does this order require shipping or will the customer pick up the order? If the answer is no, the next step is to send the order to the pick-up department, and then the order is complete. If the order does require shipping, print the invoice and label, send an

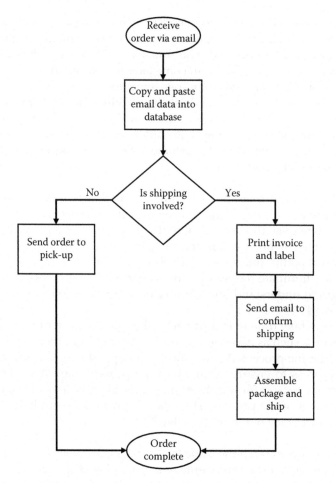

Figure 7.4 Example of a process map for receiving an order.

email confirmation, and then assemble the package and ship. Now the order is complete.

We could add more detail to the map, but these represent the most important steps for the successful completion of the order. The level of detail will often depend on the degree of specificity needed in the process to avoid unnecessary variation or redundancy. We follow the same steps for designing a new process or when a process is so broken that we must totally re-design the process. This *should-be* process map shows the ideal as we know it at this point.

The linear process map may not tell the entire story. We may need or want to delineate who does a specific step. A functional or swim lane process map, shown in Figure 7.5, can provide the information and

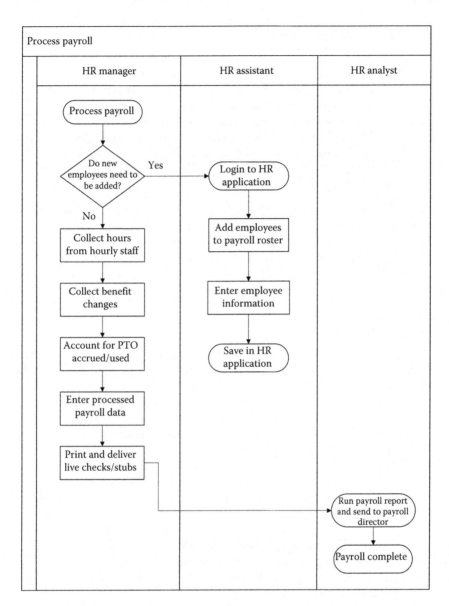

Figure 7.5 Example of a swim lane process map.

demonstrate the flow of the process across departments or functions. Different philosophies and practices exist around which type of map a team should use. In my work with teams, I keep it simple. When, and if, a functional map serves our purpose better, we can easily transfer the simple linear map into that format.

Who manages processes?

> Every organization has processes, but many do not
> understand how the processes work, do not define
> them, and do not measure their performance...
> Process Management can be seen as the single over-
> arching strategy for sustaining improvement and
> managing change. To be effective, process manage-
> ment implies ownership of processes.[5]

Leslie Steer

Processes operate in a system. We can have grand intentions about
designing and improving our processes. But, without a continual eye on
how all processes interrelate and affect each other, we may fall prey to
unintended consequences. We must constantly seek to understand how
processes work and how the handoffs from one process to another affect
outcomes.

Process management

Process Management provides a methodology that enables and empowers
people to take control of a problem (process) and develop managed solutions for that problem. Employees can feel that the management of the
process for which they have responsibility is so structured that they have
little control over any changes to improve the situation. Indeed, manage-
rial authority or existing policies and procedures may restrict the ability
to change a process.

Too few organizations develop the oversight and structures that support process design and improvement. Process management provides
governance and creates an organizational design that recognizes that
processes:

- Support the strategy of the organization
- Produce an organization's products and services
- Serve as the vehicles for meeting customer requirements and achiev-
  ing goals
- Occur horizontally and include the work of cross-functional teams
- Improve the ability of staff to do their best work

Improvement and performance excellence depends on our commitment to systematically improving critical processes, documenting them,
and monitoring their performance. In managing processes, individual
performance is only as good as the process allows.

Process owners

"Who is the process owner?" When asked this question, the response often points to a manager or frontline employee who does the work or has responsibility for the quality of the work. We mistakenly assign ownership based on proximity. The follow-up question asks, "Who makes the decisions about the resources for this process? Who can provide the funds or add personnel?"

Owning something implies control or authority. Organizations often apply the term, process owner, to the individual who monitors the performance of the process but does not have authority to add personnel or pull together cross-functional teams. The process owner has the most at stake and has authority to financially support the process. For example, the Chief Financial Officer serves as the process owner for the various financial processes. He does not manage those processes on a day-to-day basis because that responsibility may fall to a director or supervisor of an area such as accounts receivable.

This distinction becomes critical when we begin to implement improvements that require executive level support. The most disappointing consequence of process improvement efforts occurs when a team develops a solution that they cannot implement. They discover that the person, who holds the authority and purse strings, the process owner, will not provide the support or remove the barriers needed for effective implementation of the improvement. I have seen that happen far too many times.

How do you identify, design, and manage key work processes?

Critical processes align to your vision, mission, and goals of your strategy. They maximize impact on success, deliver results for goals, and ensure that the organization remains competitive. Most important, these are the processes that create value for customers, stakeholders, and the workforce. If these processes stopped, our business would suffer or grind to a halt.

Identify critical work processes

As you begin the transformation to excellence, start with the essential processes. Focus on those that have a direct impact on improving service, production, and the work of employees. Critical processes will have a clearly defined starting and ending point. Leadership and employees agree on the importance to the organization and to customers.

Start simple. Begin with effective areas that your team can make even better. Perhaps you have identified breakdowns and delays that impact

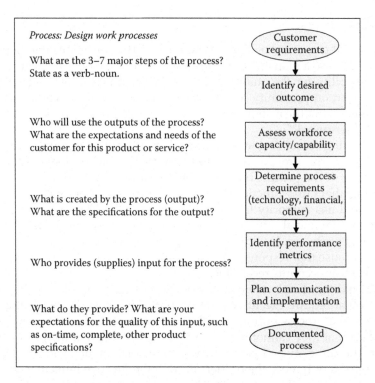

Figure 7.6 Process to identify critical work processes.

the quality of a product or service. Maybe, you have so much variation and inconsistency that the staff lack clarity on what to do or why. The process steps in Figure 7.6 should assist in answering three critical questions:

1. What are the critical success factors for the goal?
2. How will you know if you are meeting those critical success factors?
3. What processes need to be in place to address those critical success factors and achieve the goal?

Design work processes

We design processes to deliver value to our customers. We do not design processes based on our convenience. One team struggled with how to make their process more efficient. The team concluded that the problem centered on the internal customers. They just would not go from department to department to complete the process to retire. "These people just want us to do everything for them. We have a process in place. They just need to follow it. How hard can that be?"

Going from office to office took more time, created confusion, and put up barriers for staff to complete the required paperwork. The team designed the process based on their convenience—not on the needs of their customer. Rather than looking at the problem through the eyes of the retiring staff, they blamed them for the problem. No amount of memos, signs, or directions in bold would address the root cause of the problem— a cumbersome and inefficient process that failed to consider the impact on the people who mattered most.

As you examine the steps for designing work processes in Figure 7.7, keep in the forefront who the process serves. If we start the process without the customer requirements and expectations, we risk creating products and services that no one wants or needs.

Manage work processes

How do you know that processes work? How do you know if they align to your strategy? How do you know if processes have the necessary resources? Having a means to align and manage processes can help establish an infrastructure that maximizes performance and leads to expected outcomes. The benefits of intentionally and systematically managing processes include

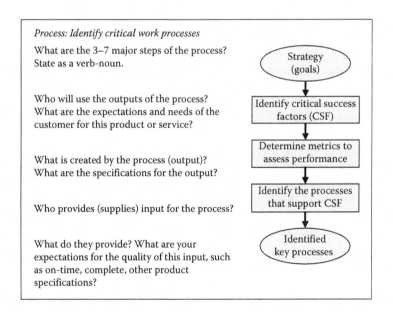

Figure 7.7 Process to design work processes.

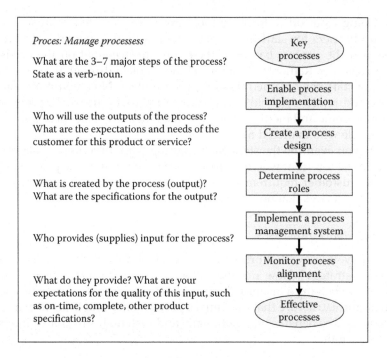

Proces: Manage processess

What are the 3–7 major steps of the process? State as a verb-noun.

Who will use the outputs of the process? What are the expectations and needs of the customer for this product or service?

What is created by the process (output)? What are the specifications for the output?

Who provides (supplies) input for the process?

What do they provide? What are your expectations for the quality of this input, such as on-time, complete, other product specifications?

Key processes

Enable process implementation

Create a process design

Determine process roles

Implement a process management system

Monitor process alignment

Effective processes

Figure 7.8 Process to manage work processes.

- Creating consistency—we minimize variation
- Increasing efficiency—we have line of sight on the handoffs that could slow the process or create waste
- Increase effectiveness—we build our capacity to achieve goals and serve customers
- Increase agility—we remain flexible and can adapt to internal and external change

Managing means we have procedures in place that support the necessary activities to align, design, implement, and monitor processes. The overall approach for managing processes reflected in Figure 7.8, provides a starting point for ensuring that all processes have a level of oversight and support. This is not about individual processes or process steps. This process creates the governance structure to effectively manage the work of the organization.

What if?

> The main thing is to keep the main thing the main thing.[6]

> **Stephen R. Covey**

What is the main thing for you or your organization? What if you and your leadership team knew which processes supported your goals? When we can answer that question, we put ourselves in a better position to make decisions around what we should do and what we should abandon.

We need to remain cognizant of the inherent variation of the system and people. Sister Mary Jean Ryan of SSM Health Care realized that "hospitals don't operate like manufacturing companies. Jobs take on the stamp of the person who does them, and changes are made by the hour, the shift, or day of the week. Tasks are passed from one person to another, often more by spoken word than by written instructions. Eventually, there is so much variation that nothing resembling a process is left.[7] Can we afford variation and inconsistency in our most critical processes? What will you choose to work on first?

We tend to hold onto systems or processes that no longer serve us or the customer. Just because we have always done it this way does not mean we should continue to do it this way. Before we can make that distinction, we must know what we do and how we do it. What worn out process can you and your team abandon?

Insights for the journey

Bryan R. Cole: Quality in education is about building a management system that drives continuous improvement of all processes based on data from customers and stakeholders.[8]

Ben Copeland: Have investment in helping people understand how to do their work.[9]

Zachary Haines: How do we identify critical processes? That's the big question. Is there a reason we are doing this and is it getting us to our rock which gets us to our goal, which helps move us toward our strategy and our mission? That's the litmus test. I'm happy when we have a meeting and discuss what's not getting us to the rock and we need to stop doing. We've only just started having those discussions in the last few years.[10]

Frankie Jackson: Our mission is to LEAD (Learn Empower Achieve Dream) in providing world-class quality K-12 systems. All processes related to that mission and the goals that have been set forth by the leadership team are all critical. The processes are interrelated and if one isn't a priority the whole system breaks down.[11]

Notes

1. G. A. Rummler and A. P. Brache, *Improving Performance: How to Manage the White Space on the Organization Chart*, 3rd ed. (San Francisco: Jossey-Bass, 2012), 43.
2. W. E. Deming. BrainyQuote.com, Xplore Inc., 2017. https://www.brainyquote.com/quotes/quotes/w/wedwardsd133510.html, accessed April 7, 2017.
3. R. Napier and R. McDaniel, *Measuring What Matters: Simplified Tools for Aligning Teams and Their Stakeholders* (Mountain View: Davies-Black Publishing, 2006), 33.
4. D. Galloway, *Mapping Work Processes* (Milwaukee: ASQ Quality Press,1994) 1, vii.
5. L. Steer, *"Process Ownership: Great Concept, But What Does It Mean?" The Human Element*, Spring 2001, Vol 18, No 2, Milwaukee: ASQ Human Development and Leadership Division as quoted in J. G. Conyers and R. Ewy, *Charting Your Course: Lessons Learned during the Journey toward Performance Excellence* (Milwaukee: American Society for Quality Press, 2004), 44.
6. S. R. Covey, *First Things First* (New York: Fireside, 1996), 13.
7. Sister M. J. Ryan, *On Becoming Exceptional: SSM Health Care's Journey to Baldrige and Beyond* (Milwaukee: ASQ Quality Press, 2007), 34.
8. B. R. Cole, EdD, *"The Role of Processes in Continuous Quality Improvement"* (lecture, Texas A&M University, College Station, Fall 2001).
9. B. Copeland, interview by author, February 20, 2017.
10. Z. Haines, interview by author, March 10, 2017.
11. F. Jackson, interview by author, March 3, 2017.

chapter eight

Design a measurement system

> Not everything that can be counted counts and not everything that counts can be counted.

Albert Einstein

The words of many industry leaders and a few unexpected others have almost become cliché since we hear them so often. We should not ignore these tidbits of advice because an element of truth and guidance exists in each one. For example, W. Edwards Deming reminds us, "In God we trust, all others must bring data," and "Without data you're just another person with an opinion." These words attributed to Peter Drucker ring especially true for me, "If you can't measure it, you can't improve it." The statistician, Edward Tufte, put it this way, "If the statistics are boring, you've got the wrong numbers." Finally, Mark Twain gets to the heart of the issue, "Data is like garbage. You'd better know what you are going to do with it before you collect it."

What does data-driven mean for excellence? The term "data-driven decision-making" has become an overused phrase in business and education, but what does it really mean and does data "drive" decisions? One of the fallacies of being "data-driven" is the assumption that data provides the answers or solutions to our problems.

Data guarantees only one thing—questions. Asking questions and pushing to uncover the root cause underlying the results requires moving beyond collecting and looking at the data. The power in the data we collect lies in how we transform it into information that will paint the picture of what otherwise might remain hidden. It is like those drawings that contain three-dimensional pictures hidden until we stare long enough to bring the object into focus.

We might think of data as that drawing and our analysis as the ability to examine and critically approach the findings. Only when we look, and look again through multiple lenses, will the picture materialize causing us to wonder why we had not seen it before. Data replaces hunches and hypotheses with facts concerning the changes needed for improvement. The data does not drive us; we are the drivers watching the road ahead, predicting the next turn, redirecting our course after a wrong turn, and looking for signs that will lead us to the place at which we hope to arrive.

Most leaders will extol the benefits of data. The paradox occurs when leaders and leadership teams fail to implement processes to regularly use the information derived from data through analysis. Believe it or not, I have sat in meetings where important decisions and discussions took place without one crumb of data. Instead, the team talked about *they, all, most, and everywhere.* Who are *they* and do we know what *they* need or want? And, just how many is *all* or *most*? Where does the problem occur? *Everywhere* tells us nothing.

We cannot underestimate the criticality of asking questions and gaining insight through tangible data and thoughtful analysis. Designing a measurement system does not happen in one day. The process is iterative, trial and error, and the results of asking many questions, such as

- What do you want to know?
- Where will you get the data?
- How do you know what you do works?
- Who owns the data?
- How do you measure your goals?
- Who will use the data and for what purpose?
- Can you act on these data?
- What drives the data?
- What else do you need to know?
- How will you share the data?

Why do we need a measurement system?

> You can have data without information, but you cannot have information without data.[1]

Daniel Keys Moran

The simplest answer to this question of why—we need to know if what we do works. We want to know the difference between good and great. Leaders use data to quantify where the organization stands relative to goals and against peers.

Quality depends on good data. It also depends on senior leaders using the data for problem solving and making decisions. An insightful school superintendent told her staff, "Don't bring me a problem without data to understand the challenge and come with a possible solution based on your analysis of the data."

The underlying concepts for excellence, systems thinking, and Deming's Theory of Profound Knowledge frame how we approach our use of data. How does the system work? Managing the system requires understanding the variation in the system that affects outcomes. The way the pieces of a system are connected (i.e., its structure) determines the outcomes it can generate. We cannot overlook two key realities regarding data and measurement:

- Changing one part of a system may affect the behavior of other parts—and the overall system—in complicated ways over time
- You can understand each part of a system in isolation and still have no idea how the overall system behaves

All systems left unattended will find their state of disorganization (the principle of entropy) or the random capacity of the system. If you have problems in a system/organization and you want to grow the system/organization, then your problems will also grow. This results from increased complexity in unmanaged systems. Put another way, if you have a problem processing purchase orders, the problem does not go away as you grow. If you do not improve or redesign the process, the problem becomes more complex over time.

A critical component of understanding the system is understanding and managing the variation in the system. The quality of our processes partly depends on the variation in the process itself and the final product or service. How much, if any, variation can exist? We cannot reduce variation or understand it unless we have the measurement system to tell us if what we do works. Knowing the extent of variation in a system includes understanding:

- Stability, capability, and predictability of processes
- Uncertainty in measurement
- Special and common causes (Chapter 2)
- Consequences of variation for all types of decision-making

We complicate this crucial component for improving our work. Finding the right measures and acting on them takes knowing our customer's requirements. We need to know if our processes deliver quality every time. The following 10 Principles of Effective Measurement[2] summarize the importance of a measurement system:

1. Define the purpose and use that will be made of the measurement
2. Emphasize customer-related measurements; be sure to include both external and internal customers
3. Focus on measurements that are useful—not just easy to collect
4. Provide for participation from all levels in both planning and implementation of measurements
5. Provide for making measurements as close in time as possible to the activities they affect

6. Provide not only concurrent indicators but also leading and lagging indictors
7. Define in advance plans for data collection and storage, analysis, and presentation of measurements
8. Seek simplicity in data recording, analysis, and presentation
9. Provide for periodic evaluations for accuracy, integrity, and usefulness of measurements
10. Realize that measurements alone cannot improve products and processes

What are the key characteristics of a measurement system?

A useful metric is both accurate (in that it measures what it says it measures) and aligned with your goals.[3]

Seth Godin

The purpose of a measurement system goes well beyond numbers and calculations. A measurement system helps us understand the outcomes, drivers, and interrelationships of our processes to make improvements throughout the system. We do not just collect data for the sake of collecting it. We want to create the knowledge necessary to make sound and effective decisions.

I can track the number of cars that drive down the street. I have data, but why did I collect the data? What will I do with it? Does it matter? If the data does not help me achieve my goal or understand a process, why do I spend precious resources collecting it? "Don't measure anything unless the data helps you make a better decision or change your actions."[4]

Measure

What makes a good measure? First, we need to define a few important terms and establish a common vocabulary as we examine measurement systems. A measure is a value that is quantified against a standard such as inches, pounds, and counts of things. The key point to remember is that there is nothing inherently good or bad with a measure. Inches are not good or bad. They are just inches.

Metric

A metric is a calculated or composite measure or quantitative indicator based upon two or more indicators or measures. Metrics might include financial ratios, emergency room wait times, student graduation rates, and volunteer retention. They help to put a variable in relation to one or more other dimensions. Metrics do provide the notion of good or bad. We now have a combination of measures that provide meaning to raw pieces of data. A metric includes three factors:

1. A measurement
2. A standard against which to compare the measurement
3. A result or outcome

Since we are dealing with language and individual's interpretations, an operational definition of metrics brings clarity and understanding. We do this before we even begin collecting data. The middle or the end of the process is too late. The operational definition defines how you will collect and record the data.

Operational definition

A precise description gives everyone involved the same understanding of what you will measure and how you plan to measure it. If we want to know the retention rate of employees, how do we define retention? What if someone works for 10 years, leaves the company, and returns? How will we address the gap in employment? Do we include part-time or contract workers in the metric? What if I take a three-month sabbatical? As you can see, depending on your perspective on retention and employees, how we use this metric could vary widely.

When we choose metrics, we want to ensure that they provide information that can be used to understand the cause, effect, and extent of the performance of our processes. The data associated with the metric must offer validity. Are we measuring the right thing? We also want reliable data. Are we measuring something consistently? Metrics should align to our strategy and processes. Is the metric relevant, and does anyone care? Finally, and most critical, can we act based on these metrics? Do we know what to do about it? Organizations possess tons of data, but do we have data to analyze a problem, determine root cause, or know what works so we can sustain the effort?

Outcome measures

Our measurement system should also include different types of measures. Outcome measures are usually high level metrics that offer insight of overall performance. These measures evaluate the results of an activity, plan, process, or program compared against projected results. In health care, outcome measures include infection and mortality rates. Educators look at student outcomes such as graduation rates and standardized test scores. Net operating profit or return on investment present common outcome measures in business or manufacturing.

Outcome measures produce lagging indicators of performance and are usually the easiest to collect. Lagging indicators make improvements during the process difficult. We cannot influence results once we have completed the work.

Process measures

In contrast, leading indicators can predict performance. Process measures allow us to measure the input of a process, the steps of the process, and the output produced by the process. These measures give us feedback along the way and an opportunity to make course corrections or improvements.

Outcome and process measures contribute to the efficacy of the measurement system. We need both, but many times, organizations focus only on outcome measures. If I want to lose 20 pounds in six months for my high school class reunion, waiting until the end of six months to step on the scale does not give me a chance to change my diet or exercise. The six months have come and gone. I can no longer change anything to get the outcome I desired.

Graduation rates give me the outcome of the teaching and learning process, but what can I do to change that outcome if I only look at that lagging indicator? What could I monitor and influence earlier in the course of a child's education? What other predictors or leading indicators would send up a red flag that students may not graduate? A few might include attendance, grades, retention in earlier grades, and even participation in extra-curricular activities.

Process measures provide evidence of the effectiveness and efficiency of our work to meet the requirements of the customer. They become the dials and flashing messages like the dashboard of our automobile. We receive a warning to change the oil before we harm the engine. Tire pressure data signals a potential flat tire. That annoying dinging reminds us that we left the key in the ignition. A good measurement system includes a balance of measures to tell us how we are doing and gives us advanced warning so we can adjust the process. For a quick sanity check, consider these questions:

- Do the metrics make sense?
- Do they measure what is important to the organization?
- Do they form a balanced set?
- Do they reinforce process monitoring and improvement?

What is a key performance indicator?

Facts do not cease to exist because they are ignored.

Aldous Huxley

Key Performance Indicators (KPIs) consists of those metrics that best indicate progress toward a desired result. At the organizational level, they provide a means to monitor the implementation of strategy. Choosing KPIs that align to the mission and goals help to determine the gap between current and desired performance. KPIs also serve as a guide to assess organizational effectiveness. Your organization's KPIs may differ from your competitor or a similar business. While we can list common indicators of success as shown in Figure 8.1, every entity must base the selection on their unique context and aspirations.

The characteristics of good KPIs do not significantly differ from those mentioned earlier. A good KPI ensures that we measure what we value as important to organizational excellence. They represent the critical few metrics upon which we can act and make decisions. KPIs align to

Examples of KPIs by Industry

Business	Education
• Cash-to-assets ratio	• Administrative expenses
• Days in accounts payable	• Average daily attendance
• Net profit growth	• Cost per meal (CPM)
• Employee engagement	• Distance learning enrollment
• Long-term debt	• Expenditures per student
• Return on equity	• Number of students per teacher
• Total cash deposits	• Transportation costs per pupil
• Labor as a percentage of cost	• Undergraduate financial aid awards
• Cycle time	• Successful course completion
• On-time orders	• Graduation rate
Health care	**Nonprofit**
• Average length of stay (ALOS)	• Gifts secured
• Cost per discharge	• Donation growth
• Turnover of clinical and non-clinical staff	• Pledge fulfilment percentage
• Risk-adjusted mortality	• Fundraising return on investment
• Number of medication errors/1000 treatments	• Donor retention rate
• FTEs per adjusted occupied bed	• Recurring gift percentage
• Cost per discharge	• Email open rate
• Admissions in-patient	• Outreach rate
• Emergency visits outpatient	• Giving capacity
• Hazardous materials usage	• Cost per dollar raised

Figure 8.1 Sample key performance indicators in education health care, nonprofits, and business.

the drivers that influence achievement of our desired outcomes. When analyzed both individually and collectively, the critical few provide an accurate, complete picture of our progress. KPIs should meet the following criteria:

- Provide critical and important data
- Be easily understood
- Be controllable by your actions
- Track actual performance change
- Align with existing data or can be "clearly" established
- Measure efficiency (timeliness, throughput, quantity, etc.) and/or effectiveness (impact, quality, contribution, etc.)

KPIs tell you the performance levels and trends of processes or systems at a given point in time. An effective performance target requires knowledge of the capabilities of the organization's critical processes or systems. Realistic stretch targets can bring a balance of pressure and support to maintain a steady increase in performance. Easy to reach targets may make us look successful but only in the short term. Eventually, the lack of rigor in the expectations we set can lead to a slow, almost imperceptible downward spiral.

The CEO of a large retail operation knew that incremental improvements built consistently over time would keep the business on a steady path toward improvement and excellence. He guided his leadership team using well-defined KPIs to develop a clear understanding of their progress and competitive position. They focused on aligning mission, goals, and action. They compared themselves against the best and remained relentless in keeping the focus on the critical few metrics.

What is the difference between a dashboard and a balanced scorecard?

> If you can't explain it simply, you don't understand it well enough.
>
> **Albert Einstein**

Do we need both? You only need what you can and will use consistently to monitor progress. If the tool you select merely sits on your desk or hangs in the hallway without any analysis or discussion, do not waste your time. The question should always center on how you will use the metrics reflected in the tool to act and make effective decisions.

The terms often create confusion, and we tend to use the terms interchangeably. But they each have a different purpose and both deserve consideration to track progress and manage performance of the organization.

A dashboard gives us real-time or near real-time data related to process metrics that give us a daily glimpse to know if what we do works. We can develop dashboards at any level of the organization to focus on short-term operational goals. We can visualize performance and know quickly if we need to adjust the process or begin looking for cause and effects.

In this age of technology, numerous applications for desktop and mobile devices can create visually attractive charts, graphs, or various types of dials to denote current levels of performance. Old fashion paper and colored markers can also create dashboards for small teams or in classrooms.

Our dashboards should resemble those in our cars, not the cockpit of a jet. Overload defeats the purpose. What data will show our progress? Can we act on the data? We don't need to make this effort complex or cumbersome.

Robert Kaplan and David Norton designed a set of measures that create a balanced scorecard to operationalize vision and strategy. The scorecard provides a balanced picture of both current and future performance through the lens of multiple perspectives and "highlights those processes that are most critical for achieving breakthrough performance for customers and stakeholders."[5] The scorecard considers performance through four perspectives: financial, customer, internal business processes, and learning and growth.

The Balanced Scorecard (BSC) offers the greatest benefit to organizational leaders when used as a management system to align vision, strategy, and key performance indicators. This implies that the BSC elevates the measurement system to a more strategic level by intentionally using it to:

- Clarify and gain consensus about strategy
- Communicate strategy throughout the organization
- Align departmental and personal goals to the strategy
- Link strategic objectives to long-term targets and annual budgets
- Identify and align strategic initiatives
- Perform periodic and systematic strategic reviews
- Obtain feedback to learn about and improve strategy[6]

Like implementing strategy, we must have a plan of action to ensure that dashboards or scorecards become dynamic processes. We start with collecting the right data and using appropriate analysis. If senior leaders do not commit to, support, or use the data, the effort can quickly become "one more thing to do" that takes my time and pulls me from my "real" work.

What story does the data tell?

> Numbers have an important story to tell.
> They rely on you to give them a clear and convinc-
> ing voice.[7]

Stephen Few

A spreadsheet or a list of calculations rarely excites and often lacks context. Data visualization breathes life into data, especially when we build context to make sense of what we see by addressing key elements. What do you intend to communicate with the data and to whom? What do you want them to do with these data? In building context, several questions can guide and direct the story you hope to tell with the data.[8]

- What background information is relevant or essential?
- Who is the audience or decision maker? What do we know about them?
- What biases does our audience have that might make them support-ive of or resistant to our message?
- What data are available that would strengthen our case? Is our audi-ence familiar with the data, or is it new?
- Where are the risks: what factors could weaken our case and do we need to proactively address them?
- What would a successful outcome look like?
- If you only had a limited amount of time or a single sentence to tell your audience what they need to know, what would you say?

The greatest challenge many of us face when presenting data is deter-mining how to set the context and tell the story briefly and succinctly. Getting to the "so what" becomes important. We must make some hard decisions regarding what data best engages the audience. We must also give them what they need to know and why they should know it before making decisions.

How important is the technique we use to illustrate data? "When we reason about quantitative evidence, certain methods for displaying data are better suited than others. Superior methods are more likely to pro-duce truthful, credible, and precise findings. The difference between an excellent analysis and a faulty one can sometimes have momentous conse-quences."[9] We have an obligation, perhaps a moral imperative, to remain honest and trustworthy in how we report data and relay the story behind it.

How do you develop and implement a measurement system?

> To measure what matters is to pay attention to the things that will have the greatest impact in achieving a purpose and refining a process.[10]

Rod Napier and Rich McDaniel

A measurement system gives the organization a common language to understand processes and systems. It becomes feedback about strategy because "success comes from having strategy become everyone's everyday job."[11] It is worth repeating again that the primary purpose of the measurement system is to measure the success of our strategy to reach our goals and fulfil our mission.

Design the measurement system

A designated team should take responsibility for developing the measurement system using the suggested process in Figure 8.2. This core group lays the groundwork for an effective approach, but, to some degree, the use of the measurement system belongs to everyone in the organization. Beware of creating a situation that gives people permission to defer measuring progress to a single unit or person.

Determine who owns the data and who will maintain responsibility for collecting, analyzing, and reporting operational process measures and key performance indicators. When these decisions remain unanswered, the measurement system will falter. In essence, no one has responsibility for it, and it will not get done.

Identify critical process measures

"Begin with the end in mind…the 'end' in this case is obtaining the information needed to effectively and efficiently plan, control, or improve."[12] Identifying process measures begins with the high-level block map of the process (SIPOC). After brainstorming possible measures of the inputs, process, and outputs, we have a good starting point for prioritizing the measures.

The team identifies the measures most critical to quality of the product or service. Before finalizing a process measure, we must validate the usefulness and appropriateness of the measures. Figure 8.3 demonstrates the key steps, and the team can test their choices by addressing these issues:

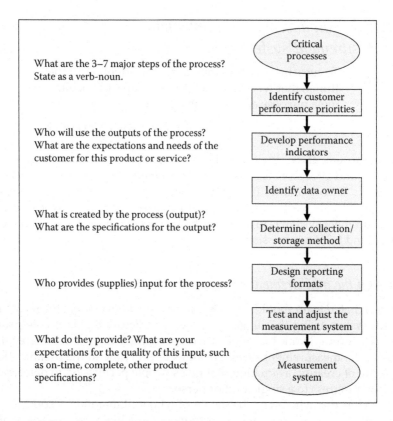

Figure 8.2 Process to design the measurement system.

- Can we do something about this?
- What will we do?
- Do we agree that the measure will help understand or improve the process?
- Do we agree that others can understand the measure?
- Do we agree that the data exists?
- Do we agree on the data owner?
- Can we gather these data with little difficulty?

Analyze data

Process depends on raw data that may be excellent, marginal, bad, or meant to deceive. We must assess data sources to determine accuracy, reliability, validity, and integrity. Our analysis depends on the quality of the data: "Invalid data nearly always produces invalid findings or 'garbage in, garbage out' as the adage goes."[13]

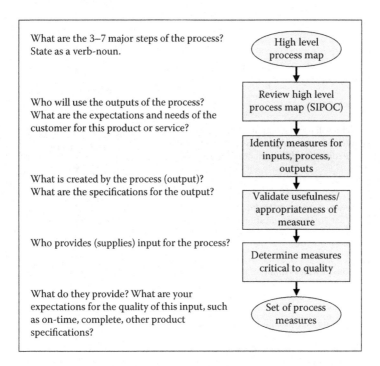

Figure 8.3 Process to identify critical process measures.

Analysis does not depend on specific software solutions although that helps. Even with the best software to slice and dice the data, without good analysis you still have only half the picture. Good analysis is a learned skill, but it requires an inquiring mind to continue to ask questions and keep digging to ferret out the real meaning behind the data. Figure 8.4 offers one approach for analyzing data. No one tool will give you every-thing you need or want to know, and, often, using different techniques can reveal new insights.

What happens when we fail to effectively analyze and interpret data? We blame people for system problems. We spend too much time on symp-toms and work on the wrong problems. With limited resources, we can-not afford to waste money on solutions that have no problem. Sometimes, we commit to actions when we should do nothing. As mentioned earlier, Deming referred to this as tinkering. Good data and analysis steers us away from needless tinkering.

What if?

Where will you choose to begin to take a closer look at what data you have and how you use it? What if you started with just one or two goals and

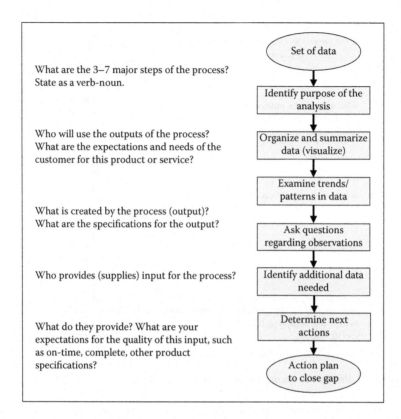

Figure 8.4 Process to analyze data.

identified the two or three measures and metrics that could provide information for action? Start small, but start with the most important results that demonstrate that you have met the needs of your customers with quality products and services.

A word of warning: If we focus the measurement system solely on monitoring and controls, we run the risk of creating mistrust. A good measurement system can build a positive environment for continuous improvement by giving us valuable feedback, creating transparency, learning about the quality of our work, and improving accountability.

As you strive to improve your use of data, I will leave this topic with a sobering thought. "Data shapes our perceptions of the real world, but sometimes we forget that data is only a partial reflection of reality." [14] Collect good data, analyze it well, but ask many questions. A participant asked me once, "What if I don't know the right questions to ask?" I gave her this advice, "Keep asking questions. Take what you learn from the answers. Ask more questions." The questions to ask regarding a set of data do not always miraculously appear. But, if we do not start somewhere, we risk missing the nugget of truth hiding among the weeds.

Insights for the journey

Genie Wilson Dillon: Effective leaders use data to drive decisions—so based upon data (from feedback, process review, competitive intelligence information, or surveys, etc.)—leaders identify opportunities for improvement.[15]

Rick Rozelle: We monitor key performance indicators, to include lagging indicators and leading indicators. To the extent possible, leading indicators are tied to the strategies in the plan and inform the executive leadership as to whether the strategies are having the proper effect. Lagging indicators are tied to the objectives in the strategic plan.[16]

JoAnn Sternke: I said it before, you measure what you treasure.[17]

Doug Waldorf: A strong and accurate measurement plan is key to monitoring progress towards goals and objectives. Typically, the more transparent the results, the stronger the alignment between organizational leadership and frontline staff. This can also be monitored through direct observations, customer feedback, and industry trends.[18]

Notes

1. D. K. Moran, BrainyQuote.com, Xplore Inc., 2017. https://www.brainyquote.com/quotes/quotes/d/danielkeys230911.html, accessed March 1, 2017.
2. J. Early and B. A. Stockhoff, "Methods and Tools: What to Use to Attain Performance Excellence," in *Juran's Quality Handbook: The Complete Guide to Performance Excellence*, 6th ed., eds. Joseph M. Juran and Joseph D. DeFeo (New York: McGraw Hill, 2010), 586.
3. S. Godin, "Avoiding False Metrics," *Seth Godin's Blog*, May 6, 2012, http://sethgodin.typepad.com/seths_blog/2012/05/avoiding-false-metrics.html.
4. S. Godin, "Analytics Without Action," *Seth Godin's Blog*, August 6, 2014, http://sethgodin.typepad.com/seths_blog/2014/08/analytics-without-action.html.
5. R. S. Kaplan and D. P. Norton, *The Balanced Scorecard: Translating Strategy to Action* (Boston: Harvard Business School Press, 1996), 11.
6. R. S. Kaplan and D. P. Norton, *The Balanced Scorecard*, 19.
7. S. Few, *Show Me the Numbers: Designing Tables and Graphs to Enlighten*, 2nd ed. (Burlingame, CA: Analytics Press, 2012), 1.
8. C. N. Knaflic, *Storytelling with Data; A Data Visualization Guide for Business Professionals* (Hoboken: John Wiley & Sons, Inc., 2015), 29.

 9. E. R. Tufte, *Visual Explanations: Images and Quantities, Evidence and Narrative,* 5th ed. (Cheshire: Graphics Press, 2002), 27.
10. R. Napier and R. McDaniel, *Measuring What Matters: Simplified Tools for Aligning Teams and Their Stakeholders* (Mountain View: Davies-Black Publishing, 2006), 24.
11. R. S. Kaplan and D. P. Norton, *The Strategy-Focused Organization*, 3.
12. J. F. Early and B. A. Stockhoff, "Accurate and Reliable Measurement Systems and Advanced Tools," in *Juran's Quality Handbook: The Complete Guide to Performance Excellence*, 6th ed., eds. J. M. Juran and J. A. DeFeo (New York: McGraw-Hill, 2010), 586.
13. B. E. Bensoussan and C. S. Fleisher, *Analysis Without Paralysis: 12 Tools to Make Better Strategic Decisions* (Upper Saddle River: Pearson Education, Inc., 2013), 17.
14. J. Harris, "Plato's Data," *The OCDQ Blog,* September 20, 2011, http://www. ocdqblog.com/home/platos-data.html.
15. G. W. Dillon, interview by author, February 13, 2017.
16. R. Rozelle, interview by author, February 22, 2017.
17. J. A. Sternke, interview by author, February 22, 2017.
18. D. Waldorf, interview by author, March 13, 2017.

chapter nine

Establish a methodology to improve

> We cannot become what we need to be by remaining what we are.[1]
>
> **Max DePree**

Achieving excellence is hard work. It requires commitment, a goal, and most important, persistence. Ask the question, "Do you want to get better at what you do?" Do you ever hear this response? "No, I would just like to remain mediocre. In fact, I think I will try to get worse at my job today."

I have never met anyone who chooses to do their worst each day. Humans come hardwired to learn, improve, aim higher, and compete with themselves and others to survive. Watch children at play. They explore. They ask *why* more than we would like. They pick things up off the ground, and to our horror, taste it. If you put up a child protection gate to keep them safe or out of mischief, they will find a way to get over the gate. Trust me. I have first-hand experience.

Whenever I begin working with a team on continuous improvement, I ask a direct question. Do you and your team want to improve and achieve excellence? All heads nod affirmative. But then, I must explain the two sides of the excellence coin. I call it the good news—bad news. The good news—I am here to help you become your best and support and coach you along the way. The bad news—to become your best will require change.

For success, excellence, and all those positive outcomes to become reality, we must prepare to make a long-term commitment to the journey.

- Are you ready to make continuous improvement a habit?
- Is your team ready for new ways to think about how and why they do the work *this way*?
- Do you and your team believe change is possible and *good news*?
- Are you prepared to look objectively at what is without blaming people?
- As the leader, will you fearlessly, persistently, and tenaciously make continuous improvement a priority?

What is continuous quality improvement?

> Quality is never an accident; it is always the result of
> intelligent effort.

> **John Ruskin**

Quality management stems from the concepts and tools of Total Quality
Management and the work of Dr. Edwards Deming. Deming believed that
analyzing and measuring processes helped identify the sources of variation that could affect the quality of a product. To explain his approach
to creating a continuous feedback loop, he built on the work of Walter
Shewhart's Plan, Do, Check, Act (PDCA) cycle.

Deming referred to it as the Plan, Do, Study, Act (PDSA) emphasizing
that the "cycle is a flow diagram for learning, and for improvement of a
product or of a process."[2] What might appear as a simple word change
had greater significance to Deming who "warned his audience that PDCA
version is not accurate because the English definition for "check" means
to "hold back".[3]

The PDSA cycle continues to evolve just as it did during Deming's lifetime. One thing remains the same—continuous study of the effectiveness
and efficiency of how we do our work with the intent to provide a quality
product to the customer. If you look, you will see the basic premise of the
PDSA cycle in some form in most improvement models.

Quality and continuous improvement demand a mindset that we can
always do better. Managing for quality also includes establishing the system
design (Chapter 2) factors that support all the activities and tasks to achieve
and maintain excellence. Stay focused because continuous improvement "is
not an add-on to our work; rather it is the way we *do* the work."[4]

We value constantly doing our best and our leaders set everyone up
for success. Leaders remove the barriers that can stifle innovation. They
design the workplace to allow for collaboration, problem solving, and looking deeply at how we do our work. Everyone from the CEO to the frontline
worker asks the question, "How do we know if what we do works?"

A set of principles guides how we manage for continuous improvement. In many ways, these reflect our values and our belief in the capacity
of people to want to do their best to meet the needs of every customer.
Embedding the following principles in the workplace sets the foundation
for excellence every day:

- The problem is in the process—the solution is in the people
- Improving quality by removing the causes of problems in the sys-
  tem inevitably leads to improved productivity
- People want to be involved and to do their jobs well

- Every person wants to feel like a valued contributor
- The person doing the job is most knowledgeable about that job
- More can be accomplished working together to improve the system than having individual contributors working around the system
- Success comes from meeting the needs and requirements of those we serve
- Unintended variation in processes can lead to unintended variation in outcomes
- We can achieve continuous improvement through small, incremental steps over time

Why do we need a process improvement team?

> For people to treat each other as teammates, they have to believe it is in their best interests to cooperate; they must be more concerned with how the system as a whole operates than with optimizing their own little piece.[5]
>
> **Brian L. Joiner**

Who should be on your process improvement team? Answering each of the following questions guides the determination of the members of the process improvement team.

The first question is most important.

- *Who owns the process?* The process owner is responsible and accountable for the process and has the ability or authority to make changes in the process.
- *Who does the work of the process?* Deming contended in his Fourteen Points that those closest to the work understood best what was wrong and how to improve the work. These individuals may only have responsibility for one step of the process, but they know that step and what makes it work and what impedes the work.
- *Who provides input to the process?* As mentioned earlier, most processes require input from other processes in the system. If the inputs of the process are faulty no amount of tinkering will improve it. You must clearly understand what is driving your process and what is needed for quality. Likewise, those who give input must know your needs and requirements for the process.
- *Who is the customer of the process?* It is all about providing the customer of your process a quality output. Do you really know the requirements and expectations of your customer, internal or external?

- *Who else does this process affect?* Are there other stakeholders who have a direct or indirect interest in your process? While they may not be your primary customer, you may consider the importance of addressing those needs or requirements.

Team selection

Do you have the right people? Take the time with your team and verify that you have the individuals who can provide the necessary knowledge to define the process. As the team begins to work on the process, you may realize that you need the expertise of someone who is not currently on your team and may not even be in your department.

Processes cut across departments and work groups in the organization. Without all stakeholders involved at some point, you could inadvertently improve your process only to create a problem, bottleneck, or delay in another process. You may even discover that your process provides input into yet another process. So, who is not here now but needs to be or may need to participate at some point during defining or analyzing this process?

Team roles

Assigning roles for each member engages everyone and expands accountability for the work of the team. The team leader oversees the work and manages the team's activities. While the group leader can also be the facilitator, the complexity of the work or dynamics of the team may require outside support.

The team scribe takes responsibility for recording and maintaining the team's work and process documents. In addition to the team leader keeping the meeting on track, a designated timekeeper monitors the effective use of time. Meetings that begin and end on time honor the team's time commitment, but it also builds a level of trust.

All team members provide knowledge of the process and participate in the discussions. Subject matter experts bring valuable in-depth knowledge of the process. Above all, do not forget the process owner!

Team meeting guidelines

Ground rules or norms provide a nonthreatening way to keep the team focused and establish how this team will work together. A parking lot chart or issue bin chart gives an outlet for deferring issues that may be important, but not relevant to the process at this point in the work. If not monitored in some way, side issues can derail a team and negatively impact team effectiveness.

Good meeting planning will assist in creating a positive and productive work environment. Even small details such as the room setup or the materials needed to complete the tasks can set the tone and expectation for collaboration.

How do we choose an effective quality improvement strategy?

> Without continual growth and progress, such words as improvement, achievement, and success have no meaning.
>
> **Benjamin Franklin**

Choose the model that best fits your organization. If you have internal expertise, select a method and adapt it to your language and the context of your organization. Do not get hung up on which one. Keep your approach simple and stick with it. Consistency and persistence will help maintain momentum. Changing methodologies every few years and following the next best thing touted as the silver bullet will hurt your efforts and credibility.

This work takes the participation of the entire organization. Continuous improvement and the tools to get there belong to everyone. Having a shepherd to guard the process keeps focus, but one person cannot single-handedly improve or transform an entire organization.

Leaders who add the responsibility for process improvement onto an already busy executive or manager, face an uphill march toward excellence. Likewise, when we place ownership in a single "quality department," senior leaders may abdicate accountability for their role in the improvement of core processes. When this occurs, no one owns the process. The process owner has the positional authority to allocate resources and make high level decisions regarding the process.

Keep things simple, but remember:

- Choose a continuous improvement approach and stick with it
- Establish and monitor metrics to evaluate your improvement efforts and outcomes
- Ensure that staff understand the methodology and the metrics
- Involve everyone

What tools do we need?

A carpenter with only a hammer, saw, screwdriver, and straight edge can build a functional cabinet. However, a carpenter with a variety of different, and sometimes, unique tools can build a distinctive and high quality product.

What differentiates a quality tool? A tool is defined as "any implement, instrument, or utensil held in the hand and used to form, shape, fasten, add to, take away from, or otherwise change something." Tools do something. Concepts are not tools because they do not *do* anything. Methodologies, or systems are not tools. Although we often refer to them as tools, they are big ideas.

Quality tools provide specific step-by-step instructions or the *how to* for the work of continuous improvement. They include such things as charts, diagrams, techniques, or methods for analysis. Quality tools help you accomplish change. Two basic sets of tools have evolved over time with two main purposes: managing and planning improvements and quality control.

Seven basic management and planning tools

1. *Affinity diagram*: Organizes brainstorming ideas into groupings and common relationships
2. *Interrelationship diagram or digraph*: Shows cause and effect relationships between factors in a complex situation
3. *Matrix diagram*: Develops a grid using an x and y axis to prioritize or analyze multiple factors
4. *Prioritization matrix*: Prioritize factors using a matrix and weighting the elements based on importance
5. *Process decision program chart*: Breaks down tasks into a hierarchy using a tree diagram
6. *Tree diagram*: A graphic organizer to break down broad categories into finer levels of detail
7. *Activity network diagram*: Sequences a set of tasks and sub-tasks that may occur in parallel and helps identify the critical path or longest sequence of events

Seven basic quality control tools

1. *Cause-and-effect diagram* (also called Ishikawa or fishbone diagram): Identifies many possible causes for an effect or problem and sorts ideas into useful categories
2. *Check sheet*: A structured, prepared form for collecting and analyzing data easily adaptable for a wide variety of purposes
3. *Control charts*: Uses a basic run chart that includes statistically determined control limits

4. *Histogram*: A graph showing frequency distributions, or how often each different value in a set of data occurs
5. *Pareto chart*: Shows on a bar graph which factors are more significant
6. *Run Chart*: Plots a series of data over time
7. *Scatter diagram*: Graphs pairs of numerical data, one variable on each axis, to look for a relationship

Additional tools for working with teams

1. *Brainstorming*: Generate ideas in a short period of time
2. *Force-field analysis*: Identifies opposing forces of change where the driving force must be stronger than the restraining forces
3. *Nominal group technique*: Brainstorm ideas in a structured format that encourages participation of all members of the group
4. *Plus/delta*: Solicits feedback on the strengths (plus) and opportunities (delta) of a meeting or situation

Your quality toolbox does not need hundreds of tools. These fourteen basic tools and the additional tools for working with your team will take you a long way in understanding how well, or if, the process works to meet customer expectations. As you master these tools, you will begin to see them as indispensable to your quality journey and an effective way to analyze process data.

Where do we start?

A methodology for continuous quality improvement

How do you start? While many frameworks exist out in the quality universe, we will use the methodology outlined in Figure 9.1. The Continuous Quality Improvement (CQI) model looks at three phases for examining a process: Phase I Understand the System, Phase II Analyze the Causes, and Phase III Improve the System. In the last phase of the model, the PDSA cycle serves as the tool to continuously learn about the capabilities of the process to deliver quality to those we serve, internally and externally.

Similar to the PDSA cycle itself, you may find many variations of this approach to continuous improvement. You will also discover similar steps and sequencing in the Six Sigma Methodology. If you are just beginning or reviving your continuous improvement efforts, this CQI model offers a basic approach that you can easily implement and adapt based on your needs.

As you begin to make decisions about where to begin improvements, your quality tools become a valuable resource. Figure 9.2 suggests where you might use the tools as you move through each process improvement step. In addition to the quality tools, a project charter outlines the essential

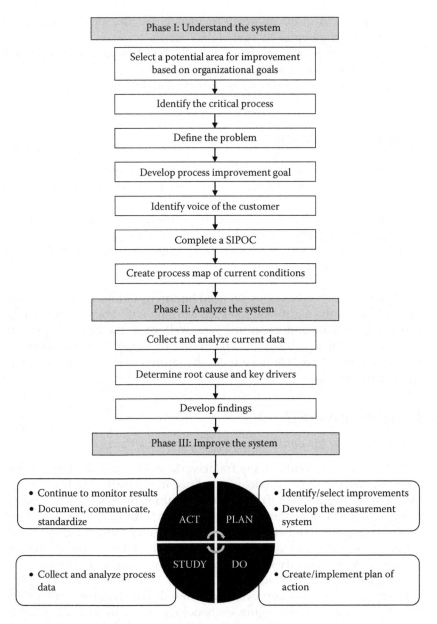

Figure 9.1 Continuous quality improvement steps.

elements of the process improvement effort including team members, alignment to organizational goals, problem and goal statements, and an action plan to capture roles and responsibilities. The charter creates a structure to communicate the improvement project and a record of key

I. Understand the system

<table>
<tr><th>Steps</th><th>Tools</th></tr>
<tr><td>• Select a potential area for improvement based on organizational goals</td><td rowspan="8">NGT
Multivoting
Affinity diagram
Brainstorming
Affinity diagram
Matrix diagram
Prioritization matrix
Tree diagram
SIPOC analysis
Flowchart
Value stream map</td></tr>
<tr><td>• Identify the critical process</td></tr>
<tr><td>• Write a problem statement</td></tr>
<tr><td>• Develop a goal statement</td></tr>
<tr><td>• Conduct a SIPOC analysis and determine the parameters of the process (including beginning and ending of the process)</td></tr>
<tr><td>• Capture voice of the customer</td></tr>
<tr><td>• Create a process flowchart/map of the current conditions</td></tr>
<tr><td>• Determine key process measures</td></tr>
</table>

II. Analyze the causes

<table>
<tr><td>• Collect and analyze current data to understand how the current process is working</td><td rowspan="5">Histogram
Pareto chart
Run chart
Check sheet
Cause and effect diagram
Force field analysis
Interrelationship digraph
Five why's</td></tr>
<tr><td>• Conduct a variation analysis</td></tr>
<tr><td>• Determine the key drivers of the process</td></tr>
<tr><td>• Determine root causes of the problem</td></tr>
<tr><td>• Develop findings</td></tr>
</table>

III. Improve the system

<table>
<tr><td>• Identify and select improvement(s)</td><td rowspan="6">Histogram
Pareto chart
Run chart
Control chart
Check sheet
Scatter plot
Flow chart
NGT
Multivoting</td></tr>
<tr><td>• Develop a measurement system</td></tr>
<tr><td>• Create and implement plans of action</td></tr>
<tr><td>• Collect and analyze process measures</td></tr>
<tr><td>• Document, communicate, and standardize the improvement</td></tr>
<tr><td>• Monitor process results to determine effectiveness of the improvement strategies</td></tr>
</table>

Figure 9.2 Tools for continuous quality improvement.

actions and deliverables assigned to members of the team. I have included a simple charter in the Appendix.

Phase I Understand the system

Select a potential area for improvement

Phase I begins by identifying an area for improvement and using a few questions to guide your thinking:

- What is not working in your area?
- What process or processes are broken?

- What are some of the most important (critical) processes in your area?
- How do you know something needs fixing?
- Do you know how those processes are doing?
- If you do know, how do you know?
- What do you look for to verify that the process is working as designed?
- What are some of the symptoms that you might notice when a process fails or is not as efficient as it could be?

Other considerations for selecting an area would include looking where you have breakdowns and delays, inefficiencies or waste, and variation in the process itself or in the implementation of the process. We do not always need to look for the bad. Sometimes we see an effective area that we could make even better.

What causes you pain? Keep in mind that the source of the pain may occur far removed from what you see and what appears as the problem. For example, do any of the following describe the symptoms you see or experience?

- Customers (internal or external receivers of the process) are unhappy
- Some things just take too long
- We throw people or money at the problem and it still does not improve
- Too many reviews and sign-offs
- High frustration while working
- The process was not done right the first time
- Processes span several departments, and there is finger-pointing and blame
- No one seems to take ownership of the total process

From this list, you may have identified several symptoms that you should address. You might identify several areas or processes that can make the most difference if we make improvements now. Remember the adage that you can only eat an elephant one bite at a time. One important question may solidify your choice: How does this process support organizational goals?

Identify the critical process
When selecting your first, or any, process to improve, you should consider if, or how, the process has a direct impact on improving service, production, and the work of employees. If it does not, you and the team may spend hours on improving something that has little to do with meeting your goals or successfully meeting the needs of the organization or customers.

Check for agreement that most employees acknowledge the importance of this process to the organization (department) and the customers of the product or services. If not, the team may not take the improvement seriously or recognize why they should take the time to improve it. Knowing if the process has a clearly defined starting and ending point may seem relatively simple. The ability to take the complexity of our work and break it down into specific processes can challenge even an experienced team. If you have doubts about where the process begins and ends, you may find the next question difficult to answer. Do you have control of the process?

In other words, can you make the decisions and monitor the changes needed for improvement? If yes, you are the process owner or his designee. Ideally, the team would include the process owner. In large organizations, the process owner may appoint a sponsor, who supports the work of the team. The process owner may also designate a champion who would bring barriers and issues back to him. The champion often serves as the life line for the team.

Ignoring the previous considerations may result in several common errors if the team selected the following:

- Process in which no one cares or believes will make a difference
- Desired solution instead of a process
- Process in transition, unstable for some reason, such as new personnel or a change in leadership

Or, you may have chosen a system to study, not a process. The human resource function of an organization has many interrelated processes with multiple hand offs across an organization. As a sub-system, improving the human resource functions will require studying many processes and their flow within and across the organization. The team will absolutely struggle because the beginning and end points occur within separate processes. Start simple, but start with something that will create value when improved.

Define the problem
In the next step, the team will identify the process problem and explain the extent of the issue and the impact of doing nothing. You may find yourself tempted to skip this step, but you will regret it later. The problem statement identifies where the problem exists, the type of improvements needed, the length of time for the current level of performance, and how the current process adversely affects the department or organization.

Writing out the problem adds clarity and purpose to the improvement project. It communicates to the team and others that a problem exists and why we need to make improvements. The problem statement does not blame people or offer excuses. When we can agree on the problem, we have taken the first step to improving the process. Figure 9.3 outlines how to write a problem statement and a goal statement with examples for each.

Develop the process improvement goal

The process improvement goal turns the problem statement into an opportunity for improvement. Taking the elements identified in the problem statement, we write a SMART goal—specific, measurable, achievable, realistic, and time bound. Using the information identified in the problem statement, we identify the direction of improvement such as increase or reduce a key indicator of the process. The team states the target for the level of future performance, the length of time for implementation, and how the improvement will benefit the organization such as cost savings, quality of cycle time, or productivity.

You will find it helpful to adhere to a verb-noun format for process names and steps. Keeping the focus on the action of the process often prevents adding process steps that may describe a document rather than the *how*. I have yet to find any process guards, but a few guidelines gleaned from experience may accelerate your learning curve.

Complete a SIPOC

How to create the SIPOC and an example appear in Chapter 3. As a reminder, the SIPOC provides a structured approach to see a process from a big picture perspective. Often referred to as a high-level block diagram, the tool often answers many critical questions about roles, responsibilities, and differences in expectations. Once you and the team have done this once or twice, you will find it can take very little time. Of all the tools you may learn in this book or elsewhere, the SIPOC helps the team to quickly organize and communicate vital information.

In addition to the process improvement goal, you must clarify the process parameters. Where does the process begin and where does it end? Do not leave this to chance or intuition—write it down.

Capture the voice of the customer

Voice of the Customer (VOC) is not an add-on or optional step when improving a process. VOC may be the most important step toward significant improvement. Often, we only examine a process from the perspective of our work, which is important, but may or may not get the end results you expected.

Understanding what the customer of the process needs or requires focuses your improvement efforts and ultimately facilitates getting to the root cause of the problem step or issue. The specifications, features, and service the customer wants and expects determines what elements from the product or service become Critical to Quality (CTQ). If we miss these,

Write the problem statement

Process	Name the process
Location	Identify where the problem exists (department, organization)
Improvement(s)	Identify the type(s) of improvements
Current performance levels	Measure the current baseline performance
Time frame	Identify the length of time for this level of performance
Impact	Determine the impact to the department or organization

Issuing purchase orders problem statement

The purchasing department's cycle time for issuing purchase orders has averaged 5 days over the past 6 months (specific dates), which has resulted in delays for purchases.

Write the goal statement

Process	Name the process
Direction	Determine the direction of the improvement
Improvement(s)	Identify the type(s) of improvements (identified in problem statement)
Improved performance levels	Determine how much the process will improve
Time frame	Identify the length of time for implementation
Impact	Determine the impact to the department or organization (savings of money, time, etc.)

Issuing purchase orders goal statement

Reduce the cycle time for issuing purchase orders to two or less days within the next quarter (specific dates), which will result in a 60%–80% decrease in cycle time.

Figure 9.3 How to write a problem statement and process improvement goal with examples.

we may have a great product or service that no one wants. Similar to the board game, Monopoly, "don't pass go" until you have stopped here.

VOC data is extremely important to understanding the effectiveness of your process. VOC may lead you to how to measure the process, but you cannot depend solely on the VOC. What do you know about your process and the expected outcomes? How will you know you are headed toward achieving that outcome?

A matrix diagram is the simplest quality tool for capturing and then analyzing VOC data. No doubt you have used this format on numerous occasions to systematically identify, analyze, or rate the correlation between a set of variables. It is a simple table, but we can glean important insights to increase our understanding of an issue or topic. The Appendix includes a VOC matrix that you can use to identify the output, the customer or customers, comments from the customer, the issue, and the specific and measurable needs or requirement expectations of the customer.

For example, the purchasing department was looking at how orders are processed and identified district staff as one of their key customers. In the lunch room, they overheard colleagues complaining about why it takes so long to receive an order. Before examining the order process, they conducted a focus group of support staff and sent out a short questionnaire to verify what they had been hearing informally. The main concern or issue in this case was the timeliness or cycle time of orders. They found out that staff needed orders processed within a week to ensure that the materials reached the department on time. If we look at this from another perspective, the purchasing department is also a customer. The input they receive from their customer (staff) is a supply order.

The purchasing department noticed more and more errors or incomplete orders and must contact the person initiating the order to get all the information needed to process their order. This may take several days. The purchasing department, as a customer, needs reliable and accurate information to process and ensure delivery within one week.

Create a process map of the current conditions
Unlike the SIPOC, the process map looks at every step in the flow of the process. Depending on the complexity of the process, the team may need to map several or all the high-level steps of the SIPOC. Recording each step of the "as-is" process will challenge the team. Teams or individuals want to map the process the way they think it happens, not how it happens. Keep asking questions until you have mapped the process as accurately as possible.

The mapping exercise will begin to reveal unnecessary approvals, too many decision points, rework or delays. The team may want to jump to a solution. Don't do it. At the end of your analysis that idea could provide the best approach, but until you complete your analysis suspend those ideas. If it helps the team, put those ideas on a chart to revisit once you have more information and data on the process.

Phase II Analyze the causes

Collect and analyze current data

You have a SIPOC, a process map, and customer requirements. The team identified data that can give them information about the process. The next step requires collection of that data. Then you hit the brick wall. We have never collected that data. We do not know how many errors cause the delay in orders. We think it takes three hours to dispense medication. We do not have a way to collect data on how many students receive special services.

If you do not have the data, or if the data would take too much time to extract from old databases, you just start somewhere. Work with what you have now, but develop a plan for how to begin to collect that data and use it as a baseline. That may mean observing and timing how long it takes to complete a step or the entire process. Without data, you can only guess at how the process currently works or does not work.

Once you have the data, use your tools to assist in the analyzes. Which tool? How many tools? That depends on the data and what you want to know.

- Do you want to know how much variation exists in the process? Does every site do it the same way and get similar results? A control chart, histogram, or a simple run chart may provide some insight.
- How often does the delay occur each week? You may only need a check list to tally the totals.
- How many calls do we receive per day over a month from each customer segment? A Pareto chart based on a simple bar graph can show who calls the most.
- Does the weight of an object affect the time it takes to deliver it? A scatter diagram will demonstrate possible correlations between two variables.

Data on a sheet of paper or in a spreadsheet will not tell you nearly as much as a graphical representation. You want to put the data in a format that allows you to ask questions about trends, relationships, and levels of performance. If you cannot ask a question, you may need to look at the data differently: "If you torture the data long enough, it will confess."[6]

Determine root cause and key drivers

Before we can begin to brainstorm solutions to improve our process, we must understand the factors affecting the outcome or effect of the process.

The fishbone diagram can help identify many possible causes of a problem. It can explain the processes of certain outcomes and facilitate brainstorming. Using a fishbone diagram to analyze the probable steps leading to a problem may also reveal potential solutions.

Another root cause tool known as the Five Whys, provides an alternative where you hunt backwards for the root cause of recurring problems. If you have children or have spent any time with a four-year old, you know exactly how this works. Why ask five and not three or ten? Generally, five questions will get you close to the root cause or at least some logical ideas. For example: The lawn mower won't start.

1. Why won't the lawn mower start? Because we don't have any gas.
2. Why don't you have any gas? Because I didn't buy any on my way back from the hardware store.
3. Why didn't you buy the gas on your way back? Because I left my credit card at home
4. Why did you leave your credit card at home? Because I didn't put it back in my wallet?
5. Why didn't you put it back in your wallet? Because this old wallet doesn't have room for more than two cards

The cause of the lawn mower not starting has nothing to do with the lawn mower. We do not need to have it serviced or buy a new one. Sometimes the cause of a problem does not occur close in time or proximity. The cost of a new wallet is much less than that of a new riding lawn mower.

The use of the interrelationship diagram or digraph clarifies the linkages between aspects of a complex problem. The key drivers responsible for the outcomes of a process may not be obvious. Taking the time to explore the relations of ideas often illuminates a critical factor that we might have otherwise overlooked.

Getting to the root cause through a cause and effect analysis pays huge dividends in saved time when we understand the true root cause of the problem. As a reminder:

- Use data to ensure that you have the right cause
- Even before you have collected data, think about how will analyze and use it
- Your cause and effect analysis may have surfaced other issues, and you can always return to this analysis to work on other causes or problems
- You may have identified many possible causes, choose the cause that, based on your data analysis, will address the problem most directly

Develop findings

You studied the data, examined potential variation, and explored the underlying causes and relation to other components of the system. Take a moment to step back. What have you learned about the current process? The myriad details, data, and analyses collected thus far can overwhelm and cause a loss of focus on the original problem of the process.

The team's work will inevitably unearth other process issues that may lure us to other critical problems that need attention. Capturing the key findings relative to this process problem provides a valuable reference of what the team has learned to help develop an effective improvement solution.

Phase III Improve the system

Now that we understand the system and have analyzed the underlying causes of the process deficiencies, we can begin to use the PDSA cycle as a tool for improving the system. A cycle contains a series of events that are regularly repeated. Continuous improvement implies that we will continuously cycle through the steps of PDSA to improve, monitor, and sustain quality process performance.

Once you have completed the initial improvements through the steps in Phase III, the people using the process may discover new ways to improve the process or find issues not previously found in the initial analysis. The team may need to start over in Phase I because of a crisis or environmental changes, but for the most part, continual cycles of improvement and refinement help ensure the stability and quality of the process.

Identify and select improvements

Teams will always want to focus on the solution and outcomes or results rather than the process. We cannot help but come with some idea about how to fix the problem. After all, we have lived with the problem for a while, and everyone knows the answer, or so they think. The process improvement team may hit on the solution but by luck rather than through analysis.

Using the PDSA, begin developing your plan on how to improve the process. Generate a variety of possible solutions before settling on one. Use brainstorming and perhaps an affinity diagram to determine the best solution. Get as many solutions as possible, no matter how crazy or unrealistic they may seem. Teams often make the mistake of only working on the first idea they come up with. This is the place to be creative, innovative, and unconventional. You might even consider combining safe and just okay solutions to create a great solution.

Consider input from customers and suppliers—ask them how the solution affects them. Include them in finding the solution. If you generated several possible solutions, a Decision Matrix can help select the best solution. When you have difficulty identifying a solution or have several from which to choose, evaluate them carefully and consider these two important questions:

- How well will the solution achieve the desired results?
- Can we successfully carry out this solution?

Before you move to implementation, brainstorm evaluation criteria. How will you know the fidelity of implementing this solution? Discuss any possible obstacles or potential problems and find ways to remove them. Continually involve stakeholders and use the feedback to continue to refine the solution. And finally, do you need additional data to determine the feasibility of the improvement?

Develop a measurement system

A measurement system (Chapter 8) becomes the dashboard by which we monitor the effectiveness and efficiency of the improvement. We will fall back to where we started if we do not plan for what data to collect, how to collect it, who collects it, and what we plan to do with the data. The data identified by the team during Phase I contributes to the system, but other measures may have surfaced along the way. You do not need hundreds; you only need those key measures and metrics that tell us how well the process works.

Create and implement plans of action

Once you have chosen a solution, plan carefully how you will carry it out. The better you plan, the easier you can prepare people for change. A good solution becomes useless if implementation fails.

The action plan guides the implementation by sequencing the major steps required to make the change. It includes what, who, and when. Implementations flounder when we underestimate the need to determine who does what before, during, and after the initial implementation of the solution. A RACI chart (See the Appendix) or some variation identifies the

- Person responsible for an action
- Person accountable for the action
- People we should consult regarding the action
- Stakeholders we should inform about the action

Establishing a timeline will keep the project from getting lost in the black hole of implementation. Accountability holds individuals responsible for specific tasks and a deadline for doing them. Time has a way of slipping past us, and we never see any results.

What are the risks to implementing this solution? A contingency plan or risk assessment can identify early what could go wrong, the likelihood of that happening, and what actions we will take if it does occur. Determine what you will do to overcome potential obstacles and plan your response to a worst-case scenario. The Appendix includes a simple tool to assess risk.

Processes, even automated ones, have a human attached to them in some way. Have you thought about the people elements that could affect success? Who will be threatened? Will the solution require learning new skills or new procedures? Will you need to test the solution before fully implementing it?

Collect and analyze process measures

Creating consistency in quality and eliminating variation drives the need for improving a process. How will you continually study and evaluate the capability of the process to deliver what the customer expects from the product or service?

Pull out your quality tool box and determine which basic quality control tool will best graphically represent the data. Consider the use of a control chart. A control chart will help you identify the amount or extent of variation acceptable to keep the process "in control." Maybe a simple run chart will expose the variation. If you do not do something with the data to help you act, your improvement efforts will suffer.

When a solution creates a large-scale change, implementing at a test site or developing a pilot serves as a means to eliminate any suspected or unforeseen problems in implementing the new process. Planning for the test or pilot includes

- Preparing an action plan for testing the solution
- Determining how to set up the pilot or test
- Identifying the results to measure the data to collect
- Assigning responsibilities
- Developing a timeline
- Involving and training those whose jobs will change by implementing the solution

Document, communicate, and standardize

WARNING: Implementation means we make a permanent change to how we do our work. That includes documenting the new process and developing a plan for storing the information and how others will access it when needed. Process maps and checklists make this easy, but if they sit in someone's desk drawer, they might as well not exist. These design issues become part of how we set up the system to support process improvement.

Use project planning, process analysis, and data collection to track and communicate the changed process. Above all else, communicate, communicate, and then communicate more. A fatal mistake occurs when we assume everyone knows about the change. Sending a single memo will not adequately prepare for any change no matter how small. A communication plan (See the Appendix) can go a long way to ensure that everyone who needs to know has the pertinent information.

To standardize or not to standardize becomes a decision worth making. If we determine that a set standard makes sense for the process and across the organization, then standardization becomes a benefit. When we standardize the *how* just to do it, we may cause resistance and, even worse, refusal to adhere to the standard.

Standards add value that ensures quality when the product or service requires a high level of consistency. When employees recognize how standards can affect a critical organizational process, they will more readily understand the need and purpose of the standard.

Standards should not

- Stifle creativity and lead to stagnation
- Interfere with a customer focus
- Add bureaucracy and red tape
- Make work inflexible or boring, especially at low-leverage process points
- Only describe the minimal acceptable outputs; be developed and not used; be etched in granite forever; be imposed from the outside; or
- Waste time

Standards should

- Make progress more visible and make it easier to track progress over time
- Capture and share lessons
- Evaluate points of consistent weakness in the process
- Evaluate how and by how much new initiatives helped to improve quality
- Help workers communicate more effectively with other functional areas
- Help workers communicate more effectively among themselves
- Be treated as living, breathing guidelines that can and *must be* constantly improved; focus on the few things in a process that truly make a difference
- Provide a *foundation* for improvements.[7]

Continue to monitor results

No one likes change, so let everyone know and understand the benefits of the improvements. Plan for what could go wrong—do not let unexpected consequences blindside the improvements. Your measurement system signals a problem or lets you know the process meets the targeted expectations. Do not forget where we began—identify who owns the new process. If no one owns the process, no one is accountable for results.

Once you have tested and introduced the improved process, ensure that the changes become routine. "New and improved" would not do much good unless everyone uses it consistently.

The success of this work depends on continually asking:

- Did you successfully implement the change?
- Did you accomplish your process objective?

The answer to these two questions determines if you need to go back in the cycle; keep in mind that continuous improvement is not linear— rather, it is iterative. What if we did not accomplish the objective or successfully implement the change? Ask more questions:

- Did you address the root cause?
- Did you develop a poor solution?
- Have you attacked the wrong cause?
- How well did you execute the plan?

Which methodology should we use?

> Quality is such an attractive banner that sometimes we think we can get away with just waving it, without doing the hard work necessary to achieve it.[8]
>
> **Miles Maguire**

What you choose does not matter. Choose a method to continually improve processes and use it consistently. You can probably find as many variations of the model just described as you can find professionals who facilitate this work across the world. Everyone wants to spin the story to make it their own. At the end of the day, you will still come back to understanding the system, analyzing causes, and improving the process.

Two other methodologies deserve attention in this discussion because of their widespread use. They each offer something a bit different.

Six Sigma

Six Sigma originated in the 1980's by Motorola, who realized that traditional statistical process control would not keep them competitive. Measuring defects per million added the level of precision required for breakthrough performance. Since that time, Six Sigma has spread among companies worldwide across various sectors.

The basic premise translates to producing defect-free products and services. Sigma represents a statistical term that measures the variation or spread of a process. Traditional quality standards sought to reduce the variation three sigmas from the process mean. A process achieves six sigmas when it does not produce more than 3.4 defects per million opportunities. A defect at six sigmas represents anything outside of the customer specifications. This level of precision has become critical for companies to effectively compete and survive in a global market.

The Six Sigma improvement methodology uses DMAIC: define, measure, analyze, improve, and control.

- *Define* the improvement project goal based on customer requirements
- *Measure* the current process and develop a measurement system to monitor progress toward the goal
- *Analyze* the current process to determine the extent of variation and the root cause of the problem
- *Improve* the process by identifying solutions to the problem
- *Control* the improved process through standardization and monitoring

Variations of this methodology guide the development of new products or services or reengineer a very broken process. Design for Six Sigma (DFSS) uses DMADV: define, measure, analyze, design, and verify. In both versions, a tollgate review verifies the completion of the steps in DMAIC or DMADV.

Lean

Lean manufacturing, lean enterprise, or simply lean employs a systematic method to eliminate waste and any nonvalue added activities in a process. Taiichi Ohno, an engineer at Toyota, often receives credit for the concepts and tools. The term first appeared in research conducted at the MIT Sloan School of Management in 1990.

We can reduce waste in almost any industry including manufacturing and a variety of service industries. Lean focuses on seven key sources of waste: transportation, inventory, motion, waiting, overprocessing, overproduction, and defects.

Lean uses the seven quality control tools and value stream mapping, which exposes areas that produce waste. Other tools and concepts associated with lean include

- 5 S: work place organization based on five Japanese words: Sort (seiri), Set in Order (seiton), Shine (seiso), Standardize (seiketsu), and Sustain (shitsuke)

- Quality Function Deployment (QFD): a structured approach to defining customer needs or requirements and translating them into specific plans to produce products to meet those needs
- Poka-Yoke: mistake-proofing that draws attention to potential human errors to eliminate product defects
- Design for manufacturing (DFM) and design for assembly (DFA): product design using lean concepts

ISO 9001-2015

The International Organization for Standardization develops and publishes over 21,000 International Standards for a plethora of industries and critical components within industries such as health care, metrology, electrical engineering resistors and capacitors. The ISO website describe standards as "world-class specifications for products, services and systems, to ensure quality, safety, and efficiency."

ISO 9001-2015 provides criteria based on principles of quality management. The strong focus on the customer ensures consistently meeting product or service requirements. This standard includes criteria related to leadership, planning, support, operation, and performance evaluation. Organization of all types, sizes and sectors can benefit from using the criteria to:

- Assess overall organizational context
- Focus on customers
- Increase efficiency
- Comply to statutory and regulatory requirements
- Identify and address risks

To meet the diverse needs of organizations, ISO has adapted quality management standards for specific industries: petroleum, medical devices, software engineering, electoral organizations at all levels of government, and local government. ISO 9000 is the only standard that offers certification.

Baldrige Excellence Framework

The Baldrige Excellence Framework and criteria served as a foundational anchor for the content of this book. I first encountered the framework and criteria in 2001. At the time, I wanted something to better understand the characteristics of a quality organization. What did that look like in education, health care, government, and business? How would you know what questions to ask to determine if you had an aligned and integrated organization? I found those questions in the criteria.

The impetus for the development of a national quality award emerged from concerns in the early 1980s about the productivity of U.S.

businesses and their ability to compete. Several organizations at the state and national level began exploring structures and funding sources for a national productivity award that would encourage American businesses to practice effective quality management and control. On August 20, 1987, President Ronald Reagan signed the Malcolm Baldrige National Quality Improvement Act of 1987 into law.

The act assigned responsibility for the award to the Department of Commerce who then gave it to one of its agencies, the National Institute for Standards and Technology (NIST). Public Law 100-107 states that a national quality award would help to improve quality and productivity by:

- Helping to stimulate American companies to improve quality and productivity for the pride of recognition while obtaining a competitive edge through increased profits.
- Recognizing the achievement of those companies that improve the quality of their goods and services and providing an example to others.
- Establishing guidelines and criteria that can be used by business, industrial, governmental, and other organizations in valuating their own quality improvement efforts.
- Providing specific guidance for other American organizations that wish to learn how to manage for high quality by making available detailed information on how winning organizations were able to change their cultures and achieve eminence.

Curt W. Reimann, the first director of the Malcolm Baldrige National Quality Award (MBNQA), spearheaded the development of the Baldrige criteria that address the key requirements for achieving performance excellence and provide the standard by which companies assess their progress and apply for the award. Reimann describes the criteria as a non-prescriptive framework that addresses the quality requirements: "It's a set of requirements that gives you considerable latitude in fashioning your own quality system" because "quality is not one thing you can write a prescription for and say that prescription fits your organization."[9] Reimann and his committee took great effort in developing a set of criteria adaptable across organizations and business sectors that would not favor practitioners or models of quality systems.

That early work resulted in an assessment instrument that can "provide a standard and specific guidance for organizations that wish to manage for high quality."[10] Every two years, quality practitioners, award recipients, and the staff at the Baldrige Performance Excellence Program review the criteria to ensure that the management concepts reflected in the questions remain current and stand as role-model activities.

A set of interrelated core values and concepts support the Baldrige Excellence Framework and represent the "embedded beliefs and behaviors found in high-performing organizations" and are the "foundation for integrating key requirements within a results-oriented framework that creates a basis for action and feedback."[11] These core values and concepts include systems perspective; visionary leadership; customer-focused excellence; valuing people; organizational learning and agility; focus on success; managing for innovation; management by fact; societal responsibility; ethics and transparency; and, delivering value and results.

The Organizational Profile sets the context of the organization by asking about their business, customers, challenges and advantages, competitors, and performance excellence strategy. The first six categories ask how leaders lead and manage organizational excellence with a focus on process efficiency and effectiveness—are we doing things the right way?

1. Leadership
2. Strategy
3. Customers
4. Measurement, analysis, and knowledge management
5. Workforce
6. Operations

The seventh category, results, asks for the evidence that those processes address their strategy and meet their challenges and goals—are we doing the right things?

As an assessment, the scoring rubrics examine the maturity of an organization on a continuum from reacting to integrated. The process rubric looks for maturity in approaches, deployment, cycles of learning, and integration. The results rubric addresses the organization's performance levels, trends, use of comparative data, and the integration of results. The value of the criteria centers on approximately 183 questions that organizations can ask themselves:

- What do we do?
- How do we do it?
- How do we know what we do works?
- What do we do if it does not work?

Today, 30 years following U.S. President Ronald Reagan signing into law the Malcolm Baldrige National Quality Award, organizations around the globe still strive toward performance excellence. Excellence across the globe includes frameworks based on the Baldrige Excellence Framework such as the European Foundation for Quality Management (EFQM) and the Asia Pacific Quality Organization (APQO) Global Performance Excellence Award. Top leaders continually seek ways to effectively and efficiently meet customer expectations; gain market share; and increase profitability.

What if?

> What is required…is a passion for quality, a mind-
> set that 'good' is not good enough. If we decide any-
> thing is good enough, we have given up on making
> it better.[12]

> **Ron Warwick**

What if you committed to creating lasting excellence one small step at
a time? The Japanese word, kaizen means "good change" "whether you
want to make a small adjustment…or transform the globe…all you have
to do is take one small—very small—step at a time."[13] If you have read
this far, you have many things to think about and consider as you focus
on transformation and excellence.

Changing an organization does not and will not happen overnight,
and I do not know of any magic wands or silver bullets that will get you
there faster. However, I do know this, "small wins fuel transformative
changes by leveraging tiny advantages into patterns that convince people
that bigger advantages are within reach."[14] Just take the first step.

Insights for the journey

Vic Figurilli: A key lesson from Baldrige is that, although the
practices of others can serve as a reference point, there is
no one-size-fits-all approach to excellence. Each organization must work to understand the principles and concepts of
Organizational Excellence and then apply them to fit its unique
characteristics and circumstances.[15]

Susan Fumo and Kelly Munson: It takes courage.[16]

Shelby Danks: Sustainability of improvement and a commitment to excellence is always a challenge when key internal
stakeholders believe it to be the new flavor or initiative of the
month. Therefore, to prevent this, it is sometimes important
to NOT communicate that "today we are staring our continuous improvement journey" or "we are now doing Baldrige"
or "we are doing this to support our performance excellence
effort." If you refer to any of your commitments as a siloed
"it," you will only be able to engage stakeholders around that
silo so long. Theory in implementation science reminds us that
there will be an implementation dip after a certain amount
of time. Instead, keep it simple and avoid heavy-handed,

over-engineered models and programs that lead to knowledge asymmetry and cut off attention to your actual organization vision at the knees.[17]

Cynthia St. John: Use of the Baldrige framework, and its full support by the CEO, was a key part of the approach to continuously improve and seek excellence. The nature of the framework is such that it inherently touches all areas necessary to achieve performance excellence. Excellence was also one of the four core organizational values.[18]

Notes

1. M. DePree, *Leadership Is an Art* (New York: Dell Publishing, 1989), 100.
2. W. E. Deming, *The New Economics for Industry, Government, and Education,* 2nd ed. (Cambridge: Massachusetts Institute of Technology Center for Advanced Educational Services, 1994), 131–132.
3. R. D. Moen and Clifford L. Norman, "Circling Back," *Quality Progress* (2010): 26.
4. Sister M. J. Ryan, *On Becoming Exceptional: SSM Health Care's Journey to Baldrige and Beyond* (Milwaukee: ASQ Quality Press, 2007), 37–38.
5. B. L. Joiner, *The Quality Yearbook Quotes on Quality* (CWL Publishing Enterprises), http://my.execpc.com/~jwoods/quotes3.htm, accessed March 30, 2017.
6. B. Dykes, "31 Essential Quotes on Analytics and Data," *Analytics Hero,* October 25, 2013, http://www.analyticshero.com/2012/10/25/31-essential-quotes-on-analytics-and-data/.
7. D. Balestracci, *Data Sanity: A Quantum Leap to Unprecedented Results,* 2nd ed. (Englewood: Medical Group Management Association, 2015), 302–303.
8. M. Maguire, "Fine Tuning Quality Tools," *Quality Progress* (October 2000), http://asq.org/quality-progress/2000/10/up-front/fine-tuning-quality-tools.html.
9. S. George, *The Baldrige Quality System: The Do-It-Yourself Way to Transform Your Business* (New York: John Wiley & Sons, Inc., 1992), 42–43.
10. R. J. Marton, *Sustaining Total Quality: Achieving Performance Excellence Using the Baldrige Award Criteria* (Rockville: Government Institutes), 4.
11. National Institute of Standards and Technology, *Baldrige Excellence Framework: A Systems Approach to Improving Your Organization's Performance* (Gaithersburg: National Institute of Standards and Technology), 40.
12. R. Warwick, *Beyond Piecemeal Improvements* (Bloomington, IN: National Education Service, 1995), 152.
13. R. Maurer, *The Spirit of Kaizen: Creating Lasting Excellence One Small Step at a Time* (New York: McGraw Hill, 2013), 7.
14. C. Duhigg, *The Power of Habit: Why We Do What We Do in Life and Business* (New York: Random House Trade Paperbacks, 2012), 112.
15. V. Figurilli, interviewed by author, February 9, 2017.
16. S. Fumo and K. Munson, interviewed by author, February 23, 2017.
17. S. Danks, PhD, interviewed by author, February 28, 2017.
18. C. St. John, PhD, interviewed by author, March 9, 2017.

section three

Sustain transformation

> I can't go back to yesterday because I was a different person then.[1]
>
> Alice in *Alice's Adventure in Wonderland*
> **(Lewis Carroll)**

Sustaining transformation assumes transformation, at some level, has occurred. What a slippery fellow transformation becomes once we set it in motion. Like stretching out a rubber band—once you let go, it reverts to its original shape. But if you stretch it long enough, the rubber band can never retain its original form. For transformation to take hold requires constant stretching, persistent pulling, creating new shapes that make it hard for the organization to return to where it began.

No other path exists. To achieve our goals and demonstrate results, everyone must adopt a mindset of continuous improvement—that good is not good enough. I have yet to find another way to get there.

"Begin with the end in mind is based on the principal that all things are created twice."[2] We must first see things in our mind's eye before we can create them in the real world. What do you see? Can you hold the rubber band of excellence taut long enough to forever change the organization?

Notes

1. L. Carroll, *Alice's Adventures in Wonderland* (Chicago: BookVirtual, 2000), 155.
2. S. R. Covey, *The Seven Habits of Highly Effective People: Restoring the Character Ethic* (New York: Simon and Schuster, 1989), 99.

chapter ten

Create a mindset for excellence

Excellence does not show up on your doorstep and ask to come in![1]

Harry Paul, John Britt, and Ed Jent

Continuous improvement first takes hold as we begin to think differently. It changes the way we think about our work and the results of that work. All of us, in one way or another, look for ways to improve our personal and professional activities. To the extent we approach this improvement purposefully and intentionally makes all the difference. Focused continuous improvement looks honestly at where we stand today and does not shirk from the gap between today and what we aim to become. A bit of reflection on our readiness to transform personally and as an organization encourages thinking and acting for excellence.

- Do you know what excellence looks like for your organization?
- Are you ready to accept that excellence requires change?
- Is your team ready to accept that excellence requires change?
- How will you ensure that continuous improvement becomes the way everyone thinks and goes about their work?
- How will you hold yourself and everyone in the organization accountable for excellence?
- How will you celebrate the small wins and the breakthrough moments?

How do you embed continuous improvement in everything you do?

Next to excellence, comes the appreciation of it.

William M. Thackeray

When we think of excellence, we think of something outstanding—of quality, the very best. But excellence is not a steady state. When we strive for excellence, we do something different. We go beyond the expected. We do not just meet the standard. We set the standard.

We need to get one thing straight—performance excellence is not a destination. If we take on that mindset, disappointment and disillusionment will invade our thinking and create a sense of hopelessness. Yes, that sounds quite harsh. Approaching this work as if we will arrive at our destination and can then relax misses the fundamental truth about continuous improvement. The word *continuous* means just that.

Excellence does not always occur with great fanfare, it often arrives in small consistent steps that have the most impact. For the athlete, it is about the striving not only the arriving. Being at the top of your game is not always enough. Sooner or later someone will do something different. We must continually stretch our thinking and practice new skills that make us better.

Performance excellence only occurs when we have ownership and intent.

We learn to see problems through the eyes of others. In understanding and identifying with the interests and needs of others, we acknowledge and respond to their view of the world—not just our own. We learn to see the same situation from a different perspective.

Do you have a culture of excellence? Assessing culture entails visibility, watching, listening, and engaging with people. You cannot mandate culture. "People *are* the culture, and the more intentional everyone in your organization (not just the founders or executives) is about what they want that culture to be, the more likely that culture will unfold in a healthy and contagious manner."[2]

Organizational culture[3] evolves through the intersection of personal and organizational values, the internal and external environment, and the structures in place within the system. Those structures consist of the system design factors discussed in Chapter 2. When leaders embed performance improvement as a key structure within the organization, a culture of excellence has fertile ground in which to grow. If continuous improvement does not become the expectation, the norm, the way we do our work and think, we have little hope of designing or sustaining efforts to improve our work or our culture.

One of the most serious impediments to quality and improvement strategies occurs when a leader walks out the door. The actions may have changed, but a culture focused on improvement never took hold. Leaders always come in and want to make their mark. Especially, when the departure of the previous leader was fraught with controversy or distrust, everything that leader accomplished becomes suspect and open for disposal. It does not matter if the prior efforts contributed to the organization's current level of success. The fact that the ousted or retiring leader initiated and supported the improvement threatens its future existence.

We have come to accept this inevitability. The impact on people seeps into the culture. The damage caused by the wholesale dismantling of hard-won change efforts erodes the confidence of those who remain in the organization. Why bother improving processes? The workers on the front line, teachers, and volunteers do not always have the full story. From their vantage point, initiatives come and go. If we do not like this one another will come soon. Eventually, resistance increases and makes even the most needed change difficult to implement.

What if leaders adopted nothing more than the processes outlined in each chapter of this book? They examined the system and asked questions to understand what works. They modeled the values across the organization and made decisions based on them. With the collaboration of leadership and input from staff and stakeholders, they built plans aligned to values and goals. People knew their value because leaders reinforced service to each other and customers. They identified critical processes and continually asked, "How do we know this works?" They chose a methodology to improve and have weaved excellence into the fabric of the organization. Leaders who do these things leave an indelible mark on people and the culture of that workplace. Those individuals who contributed to the insights for the journey in this book represent that type of leader. They create cultures of excellence wherever they go.

Alas, we cannot change human nature. We can change how we embed thinking and acting with intention and the goal of ongoing improvement. We can change the system design to build structures that create an environment that supports a continuous improvement mindset. Seven components of cultural health produce places "where people can show up authentically and powerfully

1. Shared values, vision, and purpose
2. The intention of contribution and service
3. Safety to show up, speak the truth, and take risks
4. Curiosity and vulnerability
5. Accountability and ownership
6. Reciprocity
7. Conscious measurement and rewards."[4]

The actions, words, decisions of the leader must reflect a vision for excellence, valuing excellence, and modeling excellence. People watch. They listen. They know when you live the values because "the personality of the person at the top sets the tone of the culture."[5]

Responsive leaders remain open to problems and to people's concerns. It takes risk and daring to change the rules! Leadership and performance excellence is about believing in what is possible and having the courage to break from the norm or the status quo. It is relating to change in a new way: doing what is needed to bring us closer to what we want rather than on what is impossible. We must never lose sight that continuous improvement and change represent two sides of the same coin.

We will not achieve performance excellence without understanding how change affects those we serve and our workforce. Leading with integrity, building strong teams, and developing a strong culture of collaboration give us a great place to start our transformation.

How do you communicate that change is collaborative?

> If you are leading a change effort, you need to remove the ambiguity from your vision of change.[6]
>
> **Chip Heath and Dan Heath**

Change. Where does change begin? With me—because we all want it until it visits our door. We talk about change. We blame our lack of results on change. We give credit to our success to change. We bemoan the fact that we have too much change. We throw our arms in the air and with exasperation and contend that "If those people would just change, we could get something done around here."

No matter what you think you know about change or how to manage change, one axiom regarding change will not go away. The pressure to change must be more powerful than the resistance to remain the same. How do you create just the right balance of pressure and support for change? Start with belief. "People don't believe what you tell them. They rarely believe what you show them. They often believe what their friends tell them. They always believe what they tell themselves. What leaders do: they give people stories they can tell about themselves. Stories about the future and about change."[7]

This book shares how to transform an organization one process at a time. If we substituted the word *transform* with *change*, would it mean the same? I think it does, but then, transform has a much nobler sound to it. To build the case for transformative change we must:

- Understand the system
- Lead for excellence
- Connect with those we serve
- Create a pathway for excellence
- Value people and their contributions
- Identify key work processes
- Design a measurement system
- Establish a methodology to improve
- Create a mindset for excellence

John Kotter in writing about change warns us about complacency and shares that "urgency means 'of pressing importance.' When people have a true sense of urgency, they think that action on critical issues is needed *now*, not eventually, not when it fits into a schedule. *Now* means making real progress every single day. *Critically important* means challenges that are central to success or survival, winning or losing."[8] I view excellence "of pressing importance." Our workplaces, our lives, our world needs to act on critical issues and face the challenges central to our success—indeed, our very survival.

How do you hold everyone accountable for excellence?

> Everyone thinks of changing the world, but no one thinks of changing himself.
>
> **Leo Tolstoy**

Holding others accountable must *always* begin with holding ourselves accountable. If any one action derails excellence the most, it is lack of accountability. Continuous improvement requires change, and "creating greater accountability mandates that you change the system so that it reinforces accountability at every turn."[9] The system fails to reinforce accountability when

- Leaders allow some to circumvent improvement
- The improvement does not have the support to push past the initial implementation phase
- Improvement teams improve a process but management will not implement the new process
- People are not held accountable for their actions and decisions
- Leaders at the top blame others rather than taking ownership of the problem
- Collaboration and joint ownership do not become the norm

We, individually and collectively, fail when we blame others and do not take ownership for the consequences of our actions. We must stop accepting the role of victim and rise above the circumstances we face. We all have two choices: remaining positive despite setbacks and obstacles or wallowing in anger, resentment, and pity for what we believe is unjust or unfair.

During a discussion with a colleague on change and accountability, I learned about *QBQ*. Not sure I heard him correctly, I asked him to explain.

That day, I learned a strategy that I wish I had known during a time I faced what felt like a barrage of the naysayers, complainers, and blamers.

The concept is quite simple and based on the belief that "personal accountability is about each of us holding ourselves accountable for our own thinking and behaviors and the results they produce."[10] John Miller created the *QBQ*, which stands for the question behind the question. That alone piqued my interest since I like questions. The *QBQ* can play a powerful role in redirecting our own and others negative thinking and victim syndrome that will, in turn, improve our organizations and ourselves.

I like simple, and this process has only three guidelines:[11]

1. Begin with "What" or "How" (*not* "Why," "When," or "Who")
2. Contain an "I" (*not* "they," "them," "we," or "you")
3. Focus on action

The following is a typical complaint I hear as we begin to implement new processes. "Why do we keep changing how we do this? Can't we just leave it alone for once?" Now, we redirect our thinking to create the question behind the question.

1. Change can create challenges
2. But, how can *I* contribute to this effort?
3. By going to the training next week to better understand the new system?

The next time you hear someone, or yourself, blame or complain think about the question behind the question. If nothing else, you may begin to hear your words and formulate the QBQ without realizing it. Accountability is a choice, but as leaders, we also have the responsibility to make others aware of the impact personal choice has on the work environment. It may take leaders who accept the role of "Chief Reminding Officer."[12] Repeating and modeling daily the purpose of why we do our work and why it takes everyone, every day, striving for excellence.

How do you celebrate excellence?

> The secret of joy in work is contained in one word—
> excellence. To know how to do something well is to
> enjoy it.
>
> **Pearl Buck**

How do you celebrate? Do you have a party? Do you give each other a high five? What does it mean to celebrate excellence? I have had the privilege to learn from many excellent organizations. They all have one striking thing in common. Celebration comes from a place of respect for the work of everyone in the organization. It consists of the small acts of gratitude for helping your neighbor.

One of Deming's Fourteen Points for Management states: "Drive out fear, so everyone may work effectively for the company."[13] What a simple concept! Driving out fear gets to the heart of many of the concepts that run throughout this book. It begins with a clear mission that I understand and can own. The leaders do not just talk about values, they live them. I can see every day how what they say and do reflects their commitment to excellence. I never hesitate to speak up because my ideas and thoughts have value. We continuously improve our work and take pride in the results. I understand that change is inevitable, but I look for how the change will improve our organization.

We do not need elaborate celebrations, but we do desperately want to enjoy coming to work. The small gestures of thanks for a job well done or taking the time to acknowledge the contributions of others create moments of celebration. Those moments collectively add up and excellence has a home where we can celebrate our work and each other.

What if?

> Be a yardstick of quality. Some people aren't used to
> an environment where excellence is expected.[14]

> **Steve Jobs**

What if? What if we built workplaces, one process at a time, that allowed us to do many excellent things? I often try to imagine that workplace. Fortunately, I have worked in some of those places. But I have also worked where mistrust thwarted what we could have accomplished. If you do not have the position of authority as the leader, lead from where you sit. Create excellence within your own sphere of influence. Be the change you hope to see around you. Your effort will make a difference for someone.

Insights for the journey

Genie Wilson Dillon: The commitment and expectation about improvement must be embedded in everything done or said by leaders—this needs to be an expectation—that, everyone thinks about improvement and excellence in every action and decision. Some ways this has been done include (a) commitment of staff resources to learning about improvement and methods to facilitate improvement; (b) dedicated staff that have a job focused on improvement and helping others in the organization to keep that focus and achieve excellence; (c) continual communication about this expectation from leaders—mentioning improvement and excellence in meetings, presentations, publications, on the company website, signage, and activities; (d) periodic assessments and reviews that focus on high performance—at the individual, team, department, and organizational levels; and (e) recognition for improvement accomplishments (could be at the individual, team, or department level).[15]

Brian Francis: We talk core values around here. Transforming self, transforming process, then transforming people, and then you transform your organization.[16]

Jeff Goldhorn: Good is the enemy of great. Build internal capacity at the leadership level to champion this ongoing work. It takes time and effort to build a like mind-set and then to deploy components of excellence. That is the ongoing work of our senior leaders. The leadership at the team level is critical to deploying this approach.[17]

Laura Longmire: The key challenges I face in sustaining improvements are

- Sustaining leadership
- "Silver bullets" or new initiatives used instead of using what works
- Staying focused and keeping leaders focused on the journey[18]

Doug Waldorf: There are two primary challenges that organizations face in sustaining improvements. The first is creating a culture that proactively fosters excellence, and the second is implementing effective improvements and maintaining their results. Both challenges are most appropriately addressed through strategic planning by executive leadership and strong action planning by senior leaders.[19]

Notes

1. H. Paul, J. Britt, and E. Jent. *Who Kidnapped Excellence? What Stops Us from Giving and Being Our Best?* (San Francisco: Barrett-Kohler Publishers, 2014), 112.

2. A. Cavanaugh, *Contagious Culture: Show Up, Set the Tone, and Intentionally Create an Organization That Thrives* (New York, NY: McGraw Hill Education), 102.

3. E. E. Chaffey and W. G. Tierney, *Collegiate Culture and Leadership Strategies* (New York: American Council on Education, 1988), 19.

4. A. Cavanaugh, *Contagious Culture,* 211.

5. S. Sinek, *Leaders Eat Last: Why Some Teams Pull Together and Others Don't* (New York: Portfolio/Penguin, 2014), 174.

6. C. Heath and D. Heath, *Switch,* (New York: Broadway Books, 2010), 62.

7. S. Godin, *Tribes: We Need You to Lead Us* (New York: Penguin Books, 2008), 138.

8. J. P. Kotter, *A Sense of Urgency* (Boston: Harvard Business Press, 2008), 1.

9. R. Connors, T. Smith, and C. Hickman, *The Oz Principle: Getting Results through Individual and Organizational Accountability* (New York: Penguin Group, 2004), 186.

10. J. G. Miller, *QBQ! The Question Behind the Question: Practicing Personal Accountability at Work and in Life,* (New York: G. P. Putnam & Sons, 2004), 64.

11. J. G. Miller, *QBQ!,* 106–107.

12. P. Lencioni, *The Advantage: Why Organizational Health Trumps Everything Else in Business* (San Francisco: Jossey Bass, 2012), 143.

13. W. E. Deming, *Out of the Crisis* (Cambridge: Massachusetts Institute of Technology Center for Advanced Educational Services, 1982), 23.

14. S. Jobs. BrainyQuote.com, Xplore Inc, 2017. https://www.brainyquote.com/quotes/quotes/s/stevejobs126246.html, accessed March 7, 2017.

15. G. W. Dillon, interview by author, February 13, 2017.

16. B. Francis, interview by author, February 24, 2017.

17. J. Goldhorn, interview by author, February 10, 2017.

18. L. Longmire, interview by author, March 3, 2017.

19. D. Waldorf, interview by author, March 13, 2017.

chapter eleven

Share excellence in action

The only thing to do with good advice is to pass it on. It is never of any use to oneself.

Oscar Wilde

We humans naturally seek others who share our interests, beliefs, or goals. You will find groups for runners, cyclists, musicians, and a myriad of others. Those of us who have a craving for all things related to continuous improvement, excellence, and quality get excited just talking about SIPOCs, Pareto charts, or a dynamite improvement project. We are not unlike a group of artists marveling over a new watercolor technique or golfers analyzing their stance and form. We seek those who think like us. We form tribes. As Seth Godin describes it, "a tribe is a group of people connected to one another, connected to a leader, and connected to an idea."[1]

I have found my tribe. I have made a deliberate effort to seek colleagues who share my passion regarding excellence, personally and professionally. Equally important, I want people who will challenge my current thinking. I want them to question my assumptions and, every so often, dislodge me from my own mental models. When you find these kind of people, hold on to them. They are priceless.

When I started writing this book, I knew I needed to share what I had learned from each of the contributors to "insights for the journey." These remarkable professionals come from small and large organizations, health care, education, manufacturing, business, and nonprofits. Many others have also shaped my understanding and implementation of continuous improvement. Their insights and expertise have become so embedded in how I do my work that I sometimes do not know what is my own thought or someone else's.

At a recent quality conference, I met several people who held positions like mine. We immediately began sharing, asking each other questions, taking mental notes to remember that nugget of advice. We shared openly. We learned from each other. We had a connection. I have long believed in the power of a small group of people who have a common vision and remain open to learning from others. That is the key. To get the most out of the insights shared here or anywhere, we must open our minds, listen, and allow new ideas and different perspectives to shape our thinking.

The individuals who follow shared their thoughts and experiences on transforming organizations to help you, the reader. They know from experience the transformative power of the concepts covered in this book. I asked 12 questions aligned to those key concepts. Their responses follow. I know that you will find some priceless gems to validate, expand, or nudge you along in transforming your organization one process at a time.

Insights for the journey

HOW DO YOU KEEP EVERYONE IN THE ORGANIZATION FOCUSED ON THE BIG PICTURE AND WHAT IS MOST IMPORTANT TO YOUR MISSION?

Values, mission, goals, system design, all contribute to transformation. They become the glue that keeps people and processes aligned, focused, and integrated. They establish an environment where everyone models values and mission.

RESPONSES

Genie Wilson Dillon: Ideally, employees help define the mission or at least have the opportunity to review it annually or every other year—to discuss or update it. Highly effective leaders reinforce the mission and what is most important—this is done at every single opportunity they can—every time they speak to the organization staff, or through use of publications, banners, posted signs, etc. Often the mission is incorporated into other strategies such as performance reviews, etc. This further signals the importance of everyone working to achieve the mission and keeps the focus on the big picture and what is most important.

Brian Francis: It starts on day one with our new staff when they are introduced to our mission, vision, and core values. They are also introduced by seeing our employees and watching them because the core values for us are the core values in action. People see teamwork. They see free and open communication. We have several town meetings where I highlight a core value. Every Friday we give kudos to our employees. We discuss the core values and ask employees to choose the value that reflects their greatest strength. Everything starts with our core values.

Susan Fumo and Kelly Munson: We start with our strategic plan. We train senior leaders to start with the why and put the conversation on impact.

Zachary Haines: Our mission is "generating lifelong connections to work." We help create careers for people who otherwise have challenges finding those careers, especially those with barriers and disabilities. Everyone in our organization from a cashier in one of our stores who is selling a donated item through social workers, sales folks, work ready people, they are all contributing to the mission. How do we get all our 1500 employees to know that? Constant pressure over constant time. We start on day one with onboarding. Most of our onboarding is not paperwork, rules, policies, or regulations. It is meeting the people we serve and hearing their stories. It's talking about how the engine of Goodwill works and how every single role of every single person contributes to that engine at the end of the day. There is no less or more important role. We also bring folks back in with a 60 day and annual refresher that we refer to as *revitalize*—reminding them why we're here every day.

C. Ryan Oakley: As a small organization where I have daily contact with all employees, the best way to keep everyone focused on the big picture is to lead by example. If they see me focused on the important things they will follow. If my focus is elsewhere or is inconsistent, they will tend to lose focus and be inconsistent.

Michél Patterson: Standard operating systems provide a rhythm for how you manage the organization. With 52 different clusters, we defined what the leadership rhythm needed to be. On Monday, the leadership team defined the direction for the week and reviewed how we did the week before. They would identify the focus for the next week. That focus would be cascaded into more detail into the functional meetings. We called it Meeting Monday. We focused on what we needed as an organization to achieve an objective. Leaders developed a set of comprehensive metrics that were consistent across the globe. The leadership team knew what was going on and had a forward view of what our forecast would be to set the goals.

Cynthia St. John: The mission, vision, and values (MVV) were embedded in everything the organization did (leadership decision-making, communication of decisions and strategic direction, execution of decisions/plans, etc.). There was also an overarching theme and related graphic that supported communication of the mission and vision, along with core elements of the strategic plan, which was consistently used to keep everyone on the same page. Every employee also created

Figure 11.1 Living Our Mission, Pewaukee School District. (Used with permission from JoAnn Sternke, EdD, Superintendent of Pewaukee School District.)

a concise statement about how their specific role contributed to the achievement of the larger mission and vision.

JoAnn Sternke: I think that the idea of focusing people on the big picture is the most important work we do, so we are very intentional about it. In fact, we have a process about it (shown in Figure 11.1). We call it Living Our Mission. We use four key areas to embed our mission in the culture: understand it, communicate it, nurture it, and celebrate it. We help people understand it, hire for it, and have people express it. We make it visual and something that people can see and experience. During our welcome back assembly, people talk about our mission with the person sitting next to them along with 500 others in an auditorium. They describe what it means to them or they write about the impact of the mission in their work. It's on every agenda that we do something called mission keepers. Once a week I send out 20 letters to students, faculty, or staff who have demonstrated living the mission. So, it all starts with people understanding it and expressing gratitude for it. I really believe in keeping that mission a focus. I try to really connect the dots.

Doug Waldorf: Keeping the vision in front of the team is key to organizational focus. Having employees set and reach goals is a critical part to every organization's success. Teams want their work to contribute to the organization's broader mission, and setting aligned goals makes this possible. Goal-setting is a powerful way for influencing direction and providing ongoing feedback. By clearly defining, measuring, and monitoring targets, teams are provided real-time feedback on personal and organizational performance while pushing for higher results.

HOW DO YOU MODEL YOUR ORGANIZATIONAL VALUES?

Two key themes emerge from these responses. Leaders must model and embed the core values into their day-to-day work. Engage everyone in the organization to live the values and internalize them into their day-to-day work. Leader's actions must and do speak louder than words.

Responses

Trent Beach: I have embedded times in my calendar to "get to the shop floor" to speak with and build relationships with my colleagues and reports. During these times, I am cognizant to set an example and to bring up these values in our discussions.

Ben Copeland: Being visible and setting the example. If you want people to show respect and use teamwork, then you better be showing respect and using teamwork. If you're not modeling, how can you expect others in the organization to do the same? People are watching the leadership of the organization to see if they really do what you are asking them to do.

Genie Wilson Dillon: Employees watch leaders. Effective leadership needs to consistently demonstrate the values to reinforce the understanding. Leaders' daily behaviors and actions need to model the values. This can also happen if a leader talks about it, draws attention to the values and speaks about expecting everyone to honor the values.

Susan Fumo and Kelly Munson: Kelly and I recognize the need for focus. I handed out magnifying glasses during one meeting to help people understand that they needed to focus on one thing at a time. People give lip service to that, but the schools still feel pressured to be doing many things at one time. We have a lot of different masters.

Jeff Goldhorn: Our core beliefs ultimately define our culture. We use them in the hiring process and discuss them during the interview process. We revisit them during the on-boarding process. We discuss them at meetings. We have implemented "core beliefs in action" where staff are recognized for exemplifying our core beliefs through kudos, email messages, social media, and the use of hashtags.

Zachary Haines: Every year we revisit our MVV, and this year we had a powerful discussion and came to the realization that we had seven values. We've known this for a while. You can't have seven values because nobody knows them. If you don't know them, you can't live them. We did an intensive value project. We brought everyone in the corporation to participate to determine our critical core values. Our new values are: make the difference, own it, spirit of service, and work hard—play hard. Executives didn't choose them. Staff chose the values, and staff want to live each one. We are doing a ton of work talking about our values including videos, huddles, and commenting on how staff live the values. This is our number one role as executives. If I don't walk the talk, then no one else will.

Frankie Jackson: Our team took one year to develop our values, including specific examples of behavior for each. We believe that our values are what shape our organizational culture and reflect what is important to us. Values are the essence of our organization's identity: the guiding principles for how we provide service. I strive to model these in everything I do. Our values are clearly defined on our website. All organizational recognition and focus is based on our organizational values.

Debra L. Kosarek: Focusing on everyday practices, internalizing the values, and always demonstrating that people and integrity come first. When decisions need to be made, keeping integrity in the forefront of the decision-making process, our "nonnegotiable" principles and values are modeled through deed, first and foremost. While values are also communicated through meetings, emails, and other communication tools, we walk our talk. Others will see how we model organizational values in dramatic circumstances. But, our day in and day out behaviors, when others might not be expected to notice, make the biggest statement.

Laura Longmire: Leaders have leadership report cards. Within the expected behaviors of an effective leader are approaches of how we have modeled the monthly value or any of the values. These behaviors and actions are documented and reviewed. Also at each week's "huddle," we discuss the value of the month, the organization's definition of the value, and what it means in an employee's own words. We also share "hero" stories where we witness employees demonstrating these values in their daily actions with others and our customers.

Anna Prow: First, I make sure all staff examine what those organizational values are in practice in relation to what our stated values are. If there's dissonance, I make the time to explore it with staff. Personally, I make an effort to maintain my commitment to civil society's values of equity, justice, common good so that I have integrity modelling values no matter the type of organization I'm working with.

HOW DO YOU LISTEN TO YOUR CUSTOMERS AND BUILD RELATIONSHIPS TO MEET THEIR NEEDS?

When we think of customers, we think of those external to the organization, the recipients of our products or services. We often forget that we have internal customers and building relationships and listening to their voice provide essential feedback on the quality of our work processes to meet their need.

Our first and foremost responsibility to customers is service. How do we make them feel, and how do we go above and beyond their expectations? Outstanding customer service begins with outstanding service internally—colleague to colleague.

RESPONSES

Ben Copeland: The principals are my primary customers and a small group of direct reports. I listen through any communication means possible. I may pick up the phone, call, stop by, or talk before or after a meeting—being visible. My hope is to reinforce everyone's commitment to their goals for students and teachers.

Brian Francis: Part of my modeling is making sure that staff hear that we have a very simple message, "We are here to serve." What does that look like? Well what is your role? You're an investigator, then you are here to serve the public in performing that investigation. Who are you serving? You're serving the licensee if the complaint is unfair, you're serving the consumer if the complaint is appropriate. You're serving the attorney who is going to see your work product. All of this is about service.

Susan Fumo and Kelly Munson: A lot of what we do is grassroots conversations. One of the things I guide the cabinet to do is having those critical rounding conversations on basic questions to build relationships. In a large organization like ours, the feedback that has come back is that effective, authentic communication is a concern. That includes one-on-one, written, and procedural information. We don't make assumptions. We start with the basic questions of where are we? Why are we here? Where do we need to go? Just on a day-to-day basis we have conversations regarding our processes, what we need to change and how we need to change it. As an organization, we are working on gathering the right feedback. We want to make sure we are communicating that feedback and really listening to what our staff is telling us.

Jeff Goldhorn: Customer voice is critical to our success. One of our strategic drivers is "Customer Focus: Listen to our customers, anticipate their needs, and create value that exceeds their expectations." We accomplish this through an expectation that every program or service has a process for gathering customer voice (this includes advisories, face-to-face visits, program level SWOTs, etc.). Additionally, we do an annual products and service survey, a third-party survey to gather voice on overall satisfaction with the organization and through client surveys at the end of every training that our staff facilitate.

Debra L. Kosarek: Listening to customers in ways that make sense to them is important. Gaining the input and feedback they want to give is more important than gathering the input we think we need as an organization. If the customer feels strongly enough to share it with anyone in our organization, we need to listen. It is vital that we set aside our preconceived notions of what the customer relationship should be, and view our processes through the eyes of our customers.

C. Ryan Oakley: The best way to listen to customers is to simply listen. Don't hear what you want to hear or what you think they are trying to say but listen to what they are really saying. Another way is to listen to your staff. Their feedback on customer relations and their opinions are valuable, not just because they truly have good input but also because if you want to make changes it is much easier if it is staff driven. I have found that the best way to build relationships is to make people feel good about themselves when they are with you. This is harder, but much more critical, than simply making them feel good about or impressed by you.

Anna Prow: Given my role in nonprofits, my staff are my customers, and I am in service to them as I work to strengthen social enterprises. I listen to them by truly caring about their perspectives, and soliciting their views on every relevant initiative. Over time, they know I'm for real in wanting their insights and in believing my job is to facilitate their success, and they have the trust to come to me to share their needs.

JoAnn Sternke: One of the things we try to be intentional about is making sure that we have a process or a system to listen to our customers. We talk about five levels of engagement with our community. The first one is to simply inform them at the base level. What does it mean to inform our community? We do newsletters that are strategically designed and focused on our strategic plan and mission. We send it out to everyone in our community. Information is the first level.

The next level is consultation and feedback. We survey a wide variety of customers, including student alumni one and three years out, the community, and business leaders. We talk about our workforce in a different level, but we also survey them. We want to obtain information on how well we are meeting what they say is important. We do a survey around our strategic planning process to gain the community's sense of what's important. I think that if you start with information to engage your public and then consultation and feedback closes that loop.

Involvement ensures that you start to think about how you are using more than your parent organizations or booster club. You start to get everyone involved in your organization. How can you bring people in? We talk about every parent volunteering for two hours so that they get a sense of commitment. We move that up to partnerships. This is a strategic relationship and more long term. We work hard with our senior citizens and our retirees to have an ongoing volunteer relationship. We work in reciprocal ways to involve volunteers or create partnerships with the business community.

The last level is that of ambassador for the school district and our mission. These are people who can really be your storytellers in addition to me and others. This all starts with informing, listening, involvement, partnerships, and ambassador. You strive to get people moving through that continuum so that you have more and more people tell your story. That's what is important, to have everyone living your mission.

Doug Waldorf: There are several key elements to building and maintaining customer relationships. Individual and organizational integrity are perhaps the most critical. This begins by meeting the needs of our customers. Delivering on what we say we're going to do is the foundational principle of trust and essential to long-term relationships. This includes clearly establishing and articulating what we do as an organization, how we provide our product or service and what we are committed to do should we fail to meet those expectations. It also means that we proactively manage experiences, including challenging situations, in a pleasant way. Essentially taking responsibility for what we do. Portraying a professional image reinforces our project and brand. Lastly taking calculated risk is also key with earning and keeping customers. If you want to have raving fans, sometimes you have to be raving mad.

HOW DO YOU DETERMINE THE SHORT- AND LONG-TERM GOALS FOR ORGANIZATIONAL SUCCESS?

Success of any kind requires a plan and implementation of that plan for reaching the goal. Aligning the plan to what we hope to achieve keeps us focused on our mission and goals. Organizations use different models and a variety of strategies for planning and acting on those plans. Thinking, planning, and acting strategically sets the stage for success.

RESPONSES

Trent Beach: As with many of us, some of these are handed to you. For those, I set, I first look at what is important to the organization overall and align goals to those as they relate to my area. For example, at times, financial improvements become more important than other times. When this occurs, I look at the improvements that an initiative my team can launch that would support overall financial improvements. At other times, it may be another priority. Short- and long-term horizons manifest themselves based on the goals and objectives. I work to maximize the long-term goals through change management and cultural efforts, whereas short-term goals are often more administrative actions we can implement. Success starts here but routinely requires measurement and adjustment.

Brian Francis: We are going to do four things well. We are going to issue licenses, renew licenses, verify licenses, and answer the phones to respond to our customers. Many processes exist under each of these, but groups all know which processes are critical for success. We develop functional action items that give freedom to make decisions about how to do the work.

Susan Fumo and Kelly Munson: We determine our strategies based on the goals in the strategic plan. Then, we identify the

KPIs for each strategy and two tasks that we believe will make an impact. We use the data to evaluate our progress and begin action planning. Our next step is to determine a calendar for assessments to determine progress toward hitting our goals.

Zachary Haines: We have worked hard to develop a strategic planning rhythm. That includes more than just planning. There are four phases:

1. Listening to our employees and our customers and our clients
2. Planning and building a three-year plan updated annually
3. Budgeting and building specific plans to make sure we can fund what we are doing
4. Executing the plan

We work on what we call rocks. A rock is our one-year must do, no choice, has to happen. We break those down into quarterly goals for each of the teams. These are our quarterly rocks. We're coming back and visiting those quarterly rocks every quarter and piling those together to make annual rocks. If we are doing our job right, we are getting things done on an annual basis. It is a rhythm that never stops. Like a river, it is a constant give and take as we adapt to the environment.

Laura Longmire: Long-term goals are determined through assessing current and future market/customer needs. All long-term goals must align to core mission and help drive us to achieve our vision or reset our vision. Short-term goals are set and driven by our scorecard results and targets.

Michél Patterson: In terms of strategy, I agree with what Toyota learned years ago. If you set one or two things you are going to accomplish, you will do one or two things well. When you attempt to do three to five things, you might do one well. If you set seven or ten, you may do nothing well. It's also hard to decide what you are not going to do. You must focus on only one or two things. It is the same with people. We can't change 10 behaviors at once, but we can work on one or two.

Doug Waldorf: Every organization needs goals to operate effectively. Long-term goals should be strategic in nature. Strategic plans provide organizational direction and typically focus on a five or more year period. Short-term objectives, on the other hand, usually have an operational element and include tactical action plans. Strategic plans are developed in a variety of ways including market needs assessments, past and current organizational performance, growing demand for a product or service, the competitive landscape, and advances in technology. Long-term plans should include strategic objectives while short-term priorities and action plans should include timelines, clearly defined deliverables, and robust measurement to identify success.

HOW DO YOU LISTEN TO STAFF? WHAT IMPACT DOES THIS HAVE ON THEIR COMMITMENT TO YOUR GOALS?

You cannot learn if you do not listen. Making time for dialogue may seem like a waste of time when we have so much work to do. Not taking time for conversations can spawn rumors and hearsay. Something will always fill the void if we do not listen and engage in crucial conversations.

People do not commit to goals that they don't know, believe are possible, or make a difference. You will find in the following responses examples of how these leaders promote and value input from staff. They talk about the goals and how it may affect their work. They build trust and act on their commitment. They don't make excellence an add-on. They focus on listening, integrity, and understanding that failure is not the end. Mistakes give us an opportunity to learn and make improvements.

Responses

Shelby Danks: For those leaders who are new to the journey, you don't need to announce that from now on you will be using continuous improvement principles and so you will create an employee engagement campaign and will use surveys, focus groups, and other tools. This approach is very sterile and can be interpreted by stakeholders as a "something being done TO them." Instead, just tell people you want to listen and lead better, so you've created three venues for it and would love their input. Then deploy those processes, capture their input, follow through on what they suggested, and therefore model and implement that journey. Check in on those improvements on a regular basis to make sure they are still working, seeking additional input as circumstances shift—and circumstances WILL shift. Once changes are made, stakeholders will believe the leader is doing something "with them."

Genie Wilson Dillon: Leaders spend time in staff meetings, department briefings, etc. to allow in-person time with staff to listen to their ideas, thoughts, and to understand their every-day work. They have what they call an "open-door" practice to allow any employee at any time to schedule a brief meeting to talk with the leaders. Many organizations have message broadcasts of information with the option for employees to respond and give feedback and input such as setting up call lines that allow anonymous conversation. Leaders use gatherings or meetings to ask key strategic questions or they use surveys to gather ideas, information, or concerns. Effective leaders use this information—and record this information in a way that allows them to "mine" the input for trends or common areas of concern or innovation opportunities. Often data that is gathered is "vetted" by a committee or team to determine if the ideas or input is in alignment with the organizational mission and objectives. When this approach is done systematically and with intentional integration, it often impacts goals. It conveys importance to the employees that their voice matters. They begin to see that they can influence strategic direction and work toward organizational goals.

Brian Francis: I consider these things.

- What impact does this have on their commitment to your goals?
- Is there anything in your way?
- Do we have the funding?
- What else do you need?
- What are your successes?
- What is your biggest challenge?

I think it is not just empowering staff to do the job, but empowering them to know they may make mistakes along the way. It's going to happen. I believe leaders fail their organization when they create that focus on perfection. It paralyzes folks, and it goes to finger pointing when the inevitable happens. Who's supposed to do that? Blame creates dissension, which goes against our core values of teamwork and innovation. We look at lessons learned. What can we do better the next time?

Zachary Haines: There are many methods, but the most effective is our huddle. Every single group in the organization huddles. This method allows information to go up because we have an inverted organizational chart. It is also the easiest and the fastest way for information to go down. Really listening—other methods, surveys, focus groups all pale in comparison to what we learn from those huddles. The expectation is that leaders listen in those huddles as much as they talk and bring what they have learned back up through their chain of command. This all stems from research we did years ago about how nonlinear communication works better than linear or one-way communication. We were drowning ourselves in a million emails and a million memos, websites, and newsletters. That linear communication is just not as effective as a conversation. It never will be.

Frankie Jackson: Our leadership team works in the midst of our staff every day. We have weekly T4 (Technology Transformation Tactical Team) meetings. All staff participates and reports progress toward meeting their goals. Formally, we conduct an anonymous survey to all staff, at every level. We post the results each month. I believe the impact this has on our commitment to our goals is transformational! The leadership team reviews the results and takes action in every way possible.

Laura Longmire: I listen to staff through multiple approaches. I have an open-door policy and use it daily. I do schedule daily time to walk and talk with associates. My monthly schedule has at least four sites scheduled for visits to communicate what is going on in the organization, our current results, future objectives, and time for associates to ask questions. Various meetings such as morning huddles, monthly scorecard reviews, lunch and learns, and sponsoring Kaizen events enable listening and learning from associates. Our organization conducts an annual associate satisfaction survey to listen and learn from our associates. The results of all surveys are used in the annual planning process. As a last but important listening and learning approach is the use of exit interviews.

C. Ryan Oakley: Once again, you listen by just honestly listening. You also have to show that you take their concerns and input seriously. It has a huge impact because if you have a staff that feels that their voice is heard and valued then their commitment to our goals will increase greatly.

Michél Patterson: It is imperative to set time with staff one on one. We don't just focus on activities but also on developing them. What do you want to become? A development plan is not always about training. It might be putting them in front of leaders, providing a mentor, and sometime classes or events.

Getting to the fundamental level is so important. We all want to be able to grow. How do we improve this time? Build in time for continuous improvement. Get people to work together in smaller groups so they learn to respect each other and take responsibility for parts of our team meeting agenda. I assign leads based on their expertise, such as Lean or Six Sigma. Always pay attention to the emotional side. Take the time to tell people how much we appreciate them and capture stories that help to show that.

Rick Rozelle: What impact does this have on their commitment to your goals? The project oversight process is a structured way to listen to staff as they carry out the projects required to "light up" the strategic plan. During the project team presentations to the executive leadership team, they are encouraged to discuss issues, obstacles, and risks that they are faced with in the completion of their projects. The leadership team is responsible for addressing these issues on behalf of the project team members. This helps to ensure the project team's success in implementing their piece of the strategic plan.

Doug Waldorf: There are several ways to listen to staff to gain valuable feedback on organizational health. This can be done through staff surveys, employee focus groups and forums, department meetings, one-on-one conversations with team members. These are most effective when that are regularly scheduled, communicated throughout the organization, and includes follow-up from previous interactions that require remedy. The stronger and more regular the feedback and follow-up, the greater the personal commitment to organizational goals.

HOW DO YOU DETERMINE WHICH PROCESSES ARE MOST CRITICAL TO YOUR MISSION AND GOALS?

What is the most important work you do to meet the needs of your customer? What must operate seamlessly to accomplish your mission? Begin there. Empower staff to improve what they do, but support that improvement. Identifying critical processes and documenting them gives clarity to the work we must do to reach our goals.

RESPONSES

Ben Copeland: I want my processes to be about how are you doing? What can I do to help you? What do you need?

Shelby Danks: When beginning the journey, just start by identifying the processes that need to be designed, deployed, improved, or reengineered and go do it.

Susan Fumo and Kelly Munson: Our vision is to be the first choice for all families. So, we are trying to have conscience discussions on what does that look like in measurable terms. So, the first thing we have done is to create very concise measurable understandable values. Tied to those values are very tight performance targets which will reflect the overall mission. Our next step is working with all entities to holistically determine how we are going to support children. We are sitting at the table with everybody across our five goal areas growth and achievement, engaging partnerships, climate, quality staff, and fiscal and operational stewardship. With all the key leaders, we are identifying future goals that we can all work on. We consider the return on investment. How do we assess what we are doing? How do we assess effectiveness and if the money we put into initiatives and programs impacts students and their academic growth?

Zachary Haines: Truthfully, this comes back to those quarterly rocks. We believe those annual and quarterly rocks should be driving the work we do. Any new process, change, or development should be related to those rocks. We have gotten good at asking, how does this fall into the rock? What rock is this going to move forward? It is easy to make a process for ringing out a customer at the point of sale. It's harder to write a singular process for helping a student who arrives at your school or assess someone who needs your help with their job placement. There are areas where we have well-defined processes and areas where it is more of a cadence. What has helped us is the focus on our rocks and the ability to say, "No, that does not make sense." I define a process by whether it occurs consistently every time, is reasonable, measured, improved, and has a rationale.

Debra L. Kosarek: I look at which processes are most closely and directly in contact with the customer. Which processes bring the most value, as defined by the customer? Once those processes have been identified, you can review the processes for efficiency and effectiveness, validity, and determination if performance is being effectively measured. Those processes that are the most critical are the ones bringing the greatest return on your investment of resources when you enter cycles of refinement.

Laura Longmire: We conduct a matrix analysis of strategic objectives and the alignment of core and secondary processes. The ones that have the largest impact to our mission and goals have strategic actions and measurements to track their performance and understand benchmark performance of other organizations.

C. Ryan Oakley: That's difficult to answer since I tend to be a little all over the place with this one (processes). I think that customer satisfaction fuels all. So, I would say that the processes that are most important are those that directly affect the patient/customer's experience. It should also include those that directly affect the staff's experience. You also can't ignore those that affect the bottom line.

Anna Prow: I've found that the most critical processes usually present themselves. I make a point to continuously take advantage of daily opportunities for improvement, and, given there's rarely such a thing as an isolated incident in organizations, dealing with seemingly discrete challenges often leads me to wrestling with the systems underneath them.

Rick Rozelle: CELT recommends two approaches for this. The first approach uses a capability modeling method to identify the organizational capabilities required to implement a major initiative or reform represented in the strategic plan. Capability modeling identifies the processes that must either be reengineered or developed from scratch to implement the strategic initiative or reform. The second approach uses a process classification framework to identify the high-level processes for the organization. During the semi-annual strategic planning review sessions, the executive leadership ranks the processes in the framework in terms of customer satisfaction, cycle time, quality of work, and cost of service. This ranking points the leadership to the top two or three processes that need to be improved.

Cynthia St. John: Conducting facilitated conversations with the senior leadership team resulted in a determination of which processes were considered "key" and which were considered support. The key processes were most closely linked with delivering on the organization's mission, and these processes were reviewed (in a rotating manner) during leadership meetings.

JoAnn Sternke: You think about what is job critical. We organize our organization into two big areas, the academic and the support areas. I think we know what our key work processes should be, but we often don't have the processes delineated in those areas. If you're in service, delivering high-quality customer service would be one of your key work processes. But do you really take the time to identify what that looks like for the user? When I think about a front desk secretary, what does it look like for her and what does it feel like for the parent coming in? How do you identify the tenets of that relationship? It's not rocket science. It's just making sure you know what's important.

Doug Waldorf: Process importance is a function of how much value is gained by the customer. Every organization has processes or procedures that may add little value, however, are mandated by the industry. Outside of compliance related work, activities, and processes that add the greatest value to customers should remain top of focus. This can be determined through value-stream mapping, cost and revenue results, market demand, and innovations that lead to great customer utility.

HOW DO YOU KNOW IF WHAT YOU ARE DOING WORKS? WHAT DO YOU MONITOR?

Leaders focused on transformation monitor those measures critical for achieving goals. They look at more than outcome or lagging indicators. Having measures that give constant feedback on performance allows for mid-course corrections. Dashboards, balanced scorecards, or other methods communicate progress and hold everyone accountable for results.

RESPONSES

Trent Beach: At a corporate level, I have a number of scorecards I use to determine quality and performance. These are fed by electronic data feeds as well as manually gathered inputs where it cannot be automated (yet). At the affiliated hospital level, a number of reports are available, that I can access. Select examples of KPIs include interventions to optimize/tailor medication therapies by patient, medication cost per adjusted patient admission, medication variances and associated investigative notes from root cause analyses, bar-code medication administration scan rate by hospital by unit, min/max infusion programming overrides, infusion programming reprograms, controlled substance behavioral flag reports, controlled substance diversions, productivity metric reports, and many others.

Jeff Goldhorn: We have a lot of data available to us. The challenge is determining which data sets best inform our progress toward our vision and goals. We look at month-to-month and year-to-year data comparisons. Additionally, we count on key relationships with clients to inform us. We put a lot of stock in customer voice and survey data. Lastly, we look at return clients to inform this as well: Are clients utilizing services and products from one year to the next and is the budget realizing the projected revenue?

Zachary Haines: There is no science to it. There are some KPIs that don't cascade down perfectly, but we know intuitively that they fit. Most important is getting that list of a million measures down to no more than a few per division and a few per rock. That has taken a lot of really hard work because we didn't always know which of the 30 measures were the most important. We thought we had to look at all of them. There's been statistical analysis and trial and error behind getting that list down to something we can look at on one page and have a picture. This continues to challenge us. It's just a lot of data.

If I ask someone where they stand, and they can immediately say I'm in the red on tracking, I know we have an opportunity for improvement. If I get a blank stare, we may have the wrong measure. We have chosen some that we thought would get us there and were completely wrong.

Debra L. Kosarek: I monitor processes most heavily after either their launch, or their revision/refinement. Once the process is "stable," heavy monitoring is generally not needed. Through evaluation of both in-process and end measures (leading and lagging), the process can be reviewed and modified as needed. With a complex process, monitoring the subsystems is often helpful with new and problematic processes. This type of review can shorten the time needed to perform adequate analysis and implement improvements.

Laura Longmire: Our organization uses a balanced scorecard that tracks both leading and lagging indicators. Each key performance indicator has a target that was set on either historical trends, competitive comparisons, or benchmark performance. These indicators are tracked weekly, monthly, quarterly, and annually.

C. Ryan Oakley: Several ways. I look at customer and staff feedback, specific stats that I track daily, and profitability.

Anna Prow: I often don't know that something has truly worked sustainably until I'm able to look back and see transformation. Discrete, immediate indicators only show one piece of a system and can't show how that piece works (or doesn't) as part of the whole. In fact, even if something works right away, tweaks to other parts of the system may cause enough disturbance to necessitate that I go back and do it again. As for what I monitor, I look for a process to work smoothly without my intervention or involvement.

Cynthia St. John: Key performance indicators align with strategy (to measure the success of key strategic objectives and related action plans) as well as operations (linked to key work processes). The measures were reviewed during senior leadership meetings.

JoAnn Sternke: We measure five key things: quality; customer engagement and workforce engagement and satisfaction; attention to resources, which are people, money, and time; innovation; processes, monitor how it works and what it should do. You monitor by putting it on a calendar and putting metrics to it. This is where sometimes educators, or others, say this is way too scientific. It really doesn't have to be. Are you hired by the first day of school or whatever deadline you set? That's a simple metric. Are you fully staffed, start with simple things, but set a date? We aim to be staffed by June 15 for the following year. That's a goal we set. We already have our first hires for next year because we know we get better talent if we go out early. You just start setting simple metric goals. Is hiring important, what do we want that process to look like? How do you know it works? You don't know if you don't have data.

HOW DO YOU MAKE IMPROVEMENTS TO YOUR WORK?

Organizations focused on continuous improvement use a variety of methodologies. They choose an approach that best fits their needs. The approach does not drive the improvement. Rather, intentional focus on how to do the work more efficiently and effectively to meet customer needs produces desired positive results.

Responses

Trent Beach: Sounds cliché, but literally PDCA. Measurement and adjustment is core to my daily work.

Susan Fumo and Kelly Munson: Our goal is to use SMART goals, which is PDCA process, down to the student level. We are also using it at the district level, looking for where are we in the PDSA cycle and embed it in all our decisions.

Jeff Goldhorn: Through a mind-set of continuous improvement. One of our strategic drivers is "Continuous Improvement: Create a culture of excellence through a mind-set of continuous improvement." We work to accomplish this through specific training on continuous improvement tools and models such as Baldrige performance excellence, PDSA, quality tools, an expectation that all major services and products have a WIG, lead and lag data, and customer voice.

Zachary Haines: We have a group called performance excellence (PE). They are a service organization just like information technology (IT). The organization knows that if they have an issue they are having trouble solving, they can go to PE, who will have a set of tools in their belt to solve the problem. We have Six Sigma knowledge, project management knowledge, lean kaizen, and others. Folks come to us with everything from data integrity issues to process bottlenecks, customer satisfaction issues, and error problems. We don't have a single methodology for solving those issues. We try to use the tool in our belt that best fits the problem at hand. The more tools in your belt the better off you are. For example, I used a fishbone with a team. It fleshed out so many of their issues, it was fantastic. Staff view the PE department as a trusted servant of the organization. It has taken years. There must be a high level of trust.

Laura Longmire: Most work processes are measured to several key indicators: time, quality, cost, and customer satisfaction. If there is significant rework, customer dissatisfaction, or cycle time/on-time delivery issues, the process will be looked at for opportunities for improvement. If I own the process, I will work on the process to improve it. If it is a process used by other associates, teams or departments, then a team will be formed to work on the process improvement. If a team is used, the team members are those that actively do the work of the process, while I might become the sponsor.

C. Ryan Oakley: Carefully. Personally, I've got zero problem with change. I like to burn ships and move on. If it turns out that I'm wrong, I'll go back. I have very little problem with this. Staff, however, are typically change averse. I've never really understood this but it's true. So, I, unfortunately, have to tread carefully and move more slowly than I'd like to when it comes to making improvements. I have to do a lot of explaining, get feedback, and really make sure I have the staff on board before I make a change.

Anna Prow: I assume failure at some point in every initiative, and sensitize colleagues accordingly. So, as we move forward on a project, we're all looking for opportunities to improve, and assume to learn from failures large and small.

Cynthia St. John: The system performance improvement approach was PDSA, with a heavy dose of Lean/Six Sigma woven in for more significant projects. All units were expected to demonstrate continuous improvements, particularly related to customer satisfaction and patient outcomes. Larger or cross-functional projects were chartered and received staff support from the PI department. These projects were expected to align with strategic and operational goals.

JoAnn Sternke: You don't live in the do phase. You live in the study phase of plan, do, study, act. You are intentional about putting plan, do, study, act in place. You are systematic about using time, either individually or as a department or as a team, to live in the study phase where you are analyzing the fruits of your labor. That, to me, is how you make improvements to your work. You also build a culture that wants to make improvements. You recognize that you live in a continuous improvement culture and remain intentional about committing time and reflective energy around making it better. Why didn't this work? What did work? What do we want to replicate next year? You allow time. For example, the curriculum process goes on an administrative team meeting agenda after it goes to the board so we analyze it. After the last student assessment of the year, it goes on an agenda for us to analyze. All those processes get put on an agenda annually that we analyze. Then we can all use quality tools such as a fishbone or a plus delta to understand what's working.

Doug Waldorf: Improvements are made through gathering continuous feedback on work results. As results are evaluated and compared to goals and targets, decisions are made to determine current processes or make adjustments. A standardized approach to work improvements typically yields the most consistent results. Using LEAN, SHINGO, or other improvement strategies can provide an organized and predictable method for improvement.

HOW DO YOU EMBED CONTINUOUS IMPROVEMENT AND EXCELLENCE INTO EVERY FACET OF YOUR ORGANIZATION?

Excellence does not just arrive at your door step. It takes persistence, tenacity, and never giving up. As we set up the structures and system design to support performance improvement as the norm and expectation, we gradually shape the culture. If we hire looking for individuals with a mind-set for excellence and nurture that inclination, we can develop future leaders who can carry the work forward.

Responses

Trent Beach: Our organization is multihospital and multistate. Such span is characterized by many cultural norms with which we work to bring about standardization and quality. Through efforts to enlist input and bring about congress amongst the many different affiliates, we work to improve care processes. All are asked to come to the table with excellence in mind as we plan as well as when we deploy, measure, and improve.

Ben Copeland: It is never a once and done. We are constantly looking for newer, better more efficient ways of doing things. Doing it the way we've always done it may not necessarily be the best way to do it. Letting people know they have the power to do things differently, especially when it saves time and money.

Shelby Danks: In my experience in leading the journey to excellence, I have found it is helpful to not create a separate improvement silo in the organization and call it by a quirky name, like "Continuous Improvement Journey" or "Lean Six Sigma Initiative," etc. Your organizational vision is not to be the perfect example of continuous improvement in the world. Your vision, if I may use an educational example, is to create lifelong learners who are college, career, and citizenship ready. Instead of asking employees to "do Baldrige," for example, as a leader I want them to focus on their own day-to-day work and find small ways to identify opportunities for innovation or improvement, share those with others, find solutions, implement those solutions, and evaluate how well they are working. By teaching individuals and teams to use and focus on those smaller steps instead of concerning themselves with a parallel goal of "doing continuous improvement," they are more likely to engage in the desired behaviors and sustain them over time. Small is beautiful, and a line of sight between the actual vision of the organization and those daily habits can help all stakeholders focus on the true prize—the organization's actual vision. In other words, don't create a continuous improvement initiative. Just model, teach, and lead.

Jeff Goldhorn: We have a focus on continuous improvement, as noted above. We are also part of a group of organizations that are committed to continuous improvement who meet four times throughout the year to work on performance excellence, utilizing the Baldrige Framework—South Texas Excellence Partnership (STEP). The STEP consortium includes a nonprofit, a school district, a municipality, and a community college. Each organization has committed approximately 10 people to the team and we create a common agenda to come together to work and learn with and through one another. We have brought in other collaborators along the way to learn from their experience as well.

Frankie Jackson: We embed continuous improvement and excellence as an attitude for providing service. It is a habit and a culture that is the underpinning to everything we do. We have a team that is focused on performance excellence and drives all service we provide.

Debra L. Kosarek: I view the improvements from a cross-cutting perspective. Truly meaningful and substantive improvements, sustainable ones, impact the entire organization. In viewing the improvement from this perspective, it is possible to see the opportunities for not only embedding the improvement across the organization, but engaging many staff. It also makes the improvement much easier to break into multiple projects, processes, and goals and objectives.

Laura Longmire: We use a closed-loop process based on the Shewart Cycle of PDCA. This tool is embedded in all of staff development, new employee orientation, as the last step in every action plan, and as a part of our annual performance review. As a leader, I use this continuous improvement especially in communicating and in setting expectations. "Inspecting" what your expecting is a core belief in effective leadership and communication.

C. Ryan Oakley: Engagement, education, feedback, and really paying attention. Once again, leading by example and not personally settling for the status quo when you know that you can be better. "Good is the enemy of great."

Michél Patterson: There are three key roles in any organization. Leadership has eyes for culture and what needs to change in the culture. Mid-level managers are looking with an eye for flow and the barriers to getting things through. What are the bottlenecks in the system? Colleagues at the front line have eyes for waste and always have two jobs. To do your job and improve your job. Helping people put on those glasses helps them identify what that waste is and drive it out of the system.

Anna Prow: Underneath every task is a system, or there should be. So, for the most ordinary thing, I work with staff to examine the effectiveness of the system of which it is a part. So, we find ways to improve systems as prompted by the most ordinary daily needs.

JoAnn Sternke: From the board of education to every employee, they must understand that we don't do things because we've always done them this way and the status quo is good enough. We know that we have work to do to get better so that all students meet our mission and are successful when they leave us. Everyone plays a role in continually improving. We must always strive for that excellence.

Doug Waldorf: Continuous improvement is an attitude, not an agenda item. There should be a feeling that no department or organization has "arrived." There is always greater value to be delivered to the customer. This attitude begins at the top of the organization. It is a feeling that we are never satisfied with the status quo.

HOW DO YOU COMMUNICATE WITH AND RESPOND TO STAFF, CUSTOMERS, AND STAKEHOLDERS?

Communication never ends. Just when we thought we delivered a message enough times, someone will tell us they never heard it. Dialogue, feedback loops, listening, all contribute to building trust. I want to know that you heard me and will respond to my concerns. I want to know that you always tell the truth and help me understand decisions especially those that affect my work. Of any issue, communication, or the lack of it, can sabotage the best plans and intentions. Consider communication a never-ending cycle of reinforcing messages consistently over time.

RESPONSES

Trent Beach: I have customers all around me—other departments, up line, reports, etc. I have a number of channels wherein I seek feedback and where there are not specific methods in place, I periodically ask for feedback on how well I am meeting their expectations. Key areas I seek feedback include communication (degree to which I provide the information needed, anticipate further need, and timeliness). I often do this verbally, however there are times, I will ask by email if they could "provide me with a gift" wherein I request their considered thoughts and specific actions or directions I have taken or led. I think this doubles in building relationships as it not only opens me up, but it also shows them I value the feedback and the relationship.

Genie Wilson Dillon: Good communication needs to be based upon trust, honesty, and integrity. Leaders need to be mindful that words must be supported by consistent actions and if something changes or needs to alter the original stated direction or decision, communication is necessary to encourage trust.

Susan Fumo and Kelly Munson: The goal of my team is to create and monitor and protect feedback loops throughout the district. We are continuously gathering information that will help us do better work for kids and getting it into the hands of the right people. That's my cultural message, and why I'm in your school or classroom. I'm not here to monitor or evaluate you. I am here to facilitate a conversation. I might give you information to help you keep working to get better for kids. We protect the trust level for those feedback loops. One of the things I do is run the quality peer reviews on the campuses. I want there to be a circle of trust in that process because it is a huge feedback loop. Some of that means we have the conversation that this will not be shared with anyone else. They must believe that is true and that nothing from the peer review comes back to them punitively.

Jeff Goldhorn: We have learned that it is critical that we over-communicate with staff, customers, and stakeholders. We have also learned that it is important to vary our approach. Formal meetings, informal meetings, one-on-one visits, surveys, face-to-face stand-up meetings, virtual stand-up meetings, video conference, webinar, video, print, etc.

Debra L. Kosarek: Communication is the most critical key to organizational transformation. Without transparency, staff, customers, and stakeholders may not engage with you in the needed change. People may not see the need for the changes, or attach any benefits to the change. For people to own the change(s), they must feel conversant in the changes, not only the need, cause, benefits, and importance to the level they can explain it to others. By being able to communicate, they can become champions of the transformation.

Laura Longmire: The best form of communication with associates, customers, and stakeholders is by face-to-face encounters and phone calls, followed by the second-best communication of written messages. Face to face is demonstrated through daily interactions (walk-arounds, huddles), meetings, and customer conferences. Additional face-to-face interactions are leadership roundtables, coffee with the boss, and monthly employee recognition lunches. Customers and stakeholder communications include meetings, conferences, tours, celebrations, newsletters, emails, marketing materials, surveys, and dinners. Associates absorb communications in various formats so all communications will be both oral and written. I strive to not use texting as it can be miscommunicated. The best social networking process is by phone.

C. Ryan Oakley: It depends on the circumstance. Listening is the best form of communicating. We are small, so face-to-face and one-on-one seem to get the job done in most cases.

Anna Prow: I assure them they can come to me at any time. I provide regular, specific feedback—both positive and constructive. I try to maintain compassion and humor. I try to carve time for regular meetings with my direct reports. I show them I value customer service.

JoAnn Sternke: We need to build relationships. People must hear me, see me, and not only read me. It's all about being on all those different communication channels with a common message around our mission and our work, and making sure we are connecting the dots for staff and our stakeholders. That is the most important thing we do, help connect the dots. We need to explain the why and then the what and the how.

WHAT CHALLENGES DO YOU FACE IN SUSTAINING IMPROVEMENTS AND EXCELLENCE?

The road to transformation contains obstacles, hurdles, and setbacks. How you respond to the challenges that cross your path will determine whether you approach them as the end of the road or opportunities to think more creatively to solve, mitigate, or deal with them. Preparing for change and employing a process for change can help employees understand the *why* of an effort.

Loss of key leaders always threatens the sustainability of successful change initiatives. Maintaining core values and a cohesive leadership team become nonnegotiable in times of change from within or those imposed upon us from the outside.

Responses

Genie Wilson Dillon: In most organizations, there are many priorities for focus. It may be difficult to look for opportunities or innovations when it seems more important to focus on just keeping the organization on track with everyday work. This may be a challenge and is often the case when there is staff turnover or dissatisfaction (seems to crop up in all organizations). Another challenge is when changes (based on improvements to processes) affect the data that you are monitoring. This may make it difficult to measure some areas with trend data versus a point in time. If this happens, it may create a sense of not sustaining improvements—that the targets are always changing and being adjusted. If this happens, the staff may feel demotivated and if not careful, it sends a message that nothing is ever good enough.

Brian Francis: What is the biggest challenge to culture? Change. New people come on board, and they haven't gone through our book clubs or participated in our town halls. A new leader in the middle of the legislative session may find it easy to believe that this is his core work. But their work is about having connections with employees and pulling the barriers out of the way so they can do their job. New leaders may not line up with what we are and our culture. We must make sure that the leadership team continues to reflect the core values. You can say you agree with me, but if you walk out the door and start complaining, you undermine everything we did. Once we decide which way to go—we go together.

Susan Fumo and Kelly Munson: Before an improvement comes in, we may not have a succinct process for adopting, adapting, and bringing in change. Change comes from all parts of the organization. We are working on adopting a process for change itself. We want to thoughtfully consider any change in our district through a distinct process. We can analyze and critique why we are doing it and develop some of the integral components including a communication and deployment plan. We want to ensure that even before we bring in a change, we have thought through how to sustain the change effort. We want a plan for that before we bring it in, rather than just trying to swim through it. Be more thoughtful and intentional.

Debra L. Kosarek: Improvement "fatigue" can happen when significant change is occurring in an organization. Some have the misconception that as long as change is "good," people can handle an unlimited amount. That simply isn't the case. Change, no matter how positive, requires new processes, training, behaviors, and more. These changes require creative energy and adaptation.

C. Ryan Oakley: Cash flow. Small business, small margins. I've heard before that you should practice like money didn't mean anything or like you had $10 million in the bank. You always have to think about cash flow. You will have to one way or another, and it's better to be looking at it proactively than reactively. So, that being said, you have to make smart decisions when it comes to improving. Also, getting the team to buy into your values. You have to sell them on it and that's sometimes easier said than done.

Michél Patterson: Change in leadership is the number one challenge that I have seen in keeping strategies alive and continuous improvement. Leaders come in and want to make their mark and leave behind all the work that has been done. This has been a killer to sustaining improvement. Leadership must be on board and do it from a customer perspective. You can drive out waste, but that may not be good for the customer.

Anna Prow: Inertia and default back to cultural norms, of course, threaten progress made in every initiative. I accept that setbacks will occur, and I keep my eye on the long game. Beyond culture, I also appreciate that adult learners (and busy ones at that) require repetition and diverse leadership approaches. Staff also tend to think there's a destination point to denote the end of the improvements, and I make sure to convey that improvement is never ending. Also, that transformation is never really done. Sometimes also the evangelicals I work with get discouraged, and I must remind them that change is messy and that just because we may have a setback it doesn't mean we can stop trying the improvements.

Rick Rozelle: Daily operations and interruptions get in the way of sustaining improvements.

Cynthia St. John: Focus. There are so many potential things to address or improve at any given time, and so many well-intended leaders in a large organization, the result of which can feel like an onslaught of seemingly random initiatives and requirements to employees. Efforts must be coordinated and condensed in order to have meaningful impact and not give the workforce change-lash (i.e., whiplash and weariness from the high volume and rate of change; not a term we used in the organization... just something I made up to keep it front and center in my mind).

JoAnn Sternke: I don't think we are a complacent organization, but I don't want us to become complacent. I don't want people in our organization to think that it's good enough, or because we are working hard, we need to be appreciated only for our hard work. We all need to be appreciated for our hard work and for our commitment to achieving results. Sometimes I think we can get bogged down in the, "oh we are working so hard mentality." Yes, we are, and we are doing it because it is important that we attain results for our kids. That's why we do it. So, I think the challenge is feeling overwhelmed, the lack of clarity about how much work there is, and the why of the work. All the mandates that come at us that sometime come before what we feel is important create a malaise at times. We need to prioritize and bring clarity around the work that we must do.

Doug Waldorf: In my opinion, there are two primary challenges that organizations face in sustaining improvements. The first is creating a culture that proactively fosters excellence, and the second is implementing effective improvements and maintaining their results. Both of these challenges are most appropriately addressed through strategic planning by executive leadership and strong action planning by senior leaders.

WHAT ADVICE CAN YOU OFFER OTHERS WHO WANT TO TRANSFORM THEIR ORGANIZATIONS?

The advice that follows comes from individuals who have taken on the journey of transformation and excellence. Their experiences and expertise offers us ideas, warnings, and encouragement to press on. The processes to create cultures of excellence are not hard, but they do require clear intent, consistency of purpose, and the resolve to push through roadblocks.

RESPONSES

Ben Copeland: Determine your top three things that you want to change to improve and prioritize them. Determine if there are any internal relationships between those three that if you do three it helps two. Focus on one so you can get one right before you decide to do six things at one time.

Trent Beach: For those who want to transform their organizations, I would offer tips in a few areas:

1. Encourage participation from stakeholders early in the planning process to understand their issues and address them in the plan
2. Ask stakeholders from other departments, hospitals, areas to take part in the deployment of any plans you develop. They can be a great champion once they have bought in
3. Measure, measure, and measure again. Use these metrics to know how well you are doing and make adjustments along the way
4. Persistence. Just do not give up. This is not to say that if your intended outcomes are not being achieved by the process you deployed that you don't change. Instead, this is to say if it was important when you started and it is still important along the way, chip away for small wins before giving up altogether
5. Be willing to fail. Pick yourself back up, adjust your approach, and go again

Genie Wilson Dillon: Be open to change. Gather a great group of others with you (a team approach). Be open to empowering the team to be change agents. They need the autonomy to move and change the organization to achieve the vision without the top leader guiding every step. The top leader does need to calibrate and reinforce the direction—but the team should drive it. Be clear on the vision and direction. Make sure everyone in the organization understands it and how their work or role supports it. It is helpful to identify what are the organizational strengths and what areas for focus are needed for a transformation and to reach excellence. An assessment or use of a performance excellence framework model can help give direction, but not just as a stand-alone effort. The focus on improvement and transformational change needs to be a cultural aspect of work. This means lots of modeling, reinforcement with recognition, and celebration when progress toward transformation occurs, training, communication, and perhaps benchmarking other organizations that have figured out how to transform an organization or instill a culture of improvement in their organization.

Brian Francis: The overriding mission is service. That word must mean something. It might mean taking a lot of flak and giving up personal goals. We can get so caught up on who gets credit for getting work done. When a license is awarded to a 21-year-old young single mom, who has always wanted to be a cosmetologist, what difference does it make who gets the credit? Let's get her started as quickly as possible so she can begin her dreams, so she can inspire that little girl and that little boy who look up to her. It's not about the credit. It's about serving others.

Susan Fumo and Kelly Munson: Sustainable change takes time. Pick one thing to focus on at a time. I think we often make things more complex. Sometimes the simplest route is the right route. I think there is knowledge that sustainable change takes time. If we are thoughtful about that deployment plan, and if we all work together with open-mindedness that truly is success. Transformation is a challenge. I would also give the advice that you need to weigh it out. We brought in a group of what we refer to as bright spots, those committed and energetic people in the organization. We remind ourselves that even bright spots have dim days. It is not a Pollyanna experience. There are often dim days that must be sustained through. Change is painful, takes times, and brings uncomfortable moments. Consider getting as many to the table to talk about the change and get as much feedback as possible across stakeholder groups. Because dim days will come, carve out time to celebrate. We are amazed to see the power of transparency. The best dialogue occurs when we share openly and remain honest about where we are.

Jeff Goldhorn: Use the Baldrige Framework as a starting point. Identify a team who will serve as champions for this work. Write an organizational profile early on in the process. Don't write one to submit for review, write one for the value that the process provides. It is hard work, but it also brings the team that is working on it together in a way that results in a common vision and approach. Do not make it about an award. Team up with other like-minded organizations and people who can support you and with whom you can learn with and through.

Zachary Haines: Do not undertake the journey alone. Cultural change and a transforming journey take true buy-in not only from the executive team but from every level in the organization. There must be an understanding up front that it is going to take time, energy, and work. Change is going to happen and that change isn't always comfortable. When I first started in this field, I tried to be a single point of change. I learned quickly that it should be a group push. We need to evaluate and assign a value to everything we do. Change is not always perceived as a good thing. But when you talk in terms of hours saved, or lives helped, or wait time reduced or tangible improvement, suddenly, the work carries more value. One of our rules is that we don't engage with a team unless we know that we will be able to talk about value after the engagement. Remember that transforming an organization takes pressure over time. Underline time. We are 10 years into it, and that's all I can tell you. We're not 10 years closer. We're not 10 years away. We're just 10 years in, and we are way better than we were. We've taken steps forward and backward. It's just pressure over time. You must be willing to have a relentless push. Once folks accept that, it goes a lot easier.

Frankie Jackson: Transformation occurs every moment of the day. You never reach a state of total transformation. It is continuous and requires constant care and attention. The journey is in the approach and deployment with refinement—and of course have fun along the way.

Debra L. Kosarek: Be patient and kind in your efforts. Know that you are unsettling the work environment in the organization. Look for opportunities for others to own the changes. Look for ways to tie the changes to needs others have cited. Always look for opportunities to credit the employees with the changes.

Laura Longmire: My best advice is understanding it's a journey. Start with assessment of where you are. Develop an understanding and agreement of where you want to go. Measure, track progress, and ensure accountability with buy-in at all levels. Above all leaders must role model and constantly communicate what is the vision and how are we going to get there. Communication must be transparent. Oh, yes, align and link associate recognition and rewards to the goals and objectives.

C. Ryan Oakley: It's been a long journey, as I guess it should. Ups and downs, more ups than downs, thankfully. But it never ends. Complacency is not an option. There's always something more to be done. You have to enjoy the journey, enjoy the process.

Michél Patterson: There is a lean concept that you turn the organizational chart upside down. The people who are at the value stream providing value for your customers, which is the front line, they become the purpose for your leadership. What do we need to do to make it as easy as possible to make value for customers? That is our job. Have many listening posts for customers and colleagues. Create experience for colleagues. I see companies who view people as a resource that is no different than a widget. In fact, they try to get rid of people, rather than helping them create more value and grow them. When you treat people as a cost to your organization, you start treating them as widgets. This creates a downward spiral because staff translate that to customers as well. Provide simple tools and encouragement to teach leaders and include a coaching model to drive culture.

Anna Prow: Social enterprise leadership sometimes does not have the capacity or expertise to "own the problems" and create custom solutions to organizational challenges—or to manage the changes necessary to surmount them sustainably. And, for a number of reasons, social enterprise business and planning systems often aren't resilient enough to accommodate organizational evolution and keep program work and business operations in coordination. In some cases, to drive organizational transformation, social enterprises should engage a capable and experienced change and learning agent at the inflection points in their growth to identify organizational challenges and own the resolution of those challenges in the context of organizational strengths and limitations.

Rick Rozelle: Focus the work on the right things—and the right things are defined by the strategic plan. Use an executive project oversight process that requires a weekly meeting of the executive leadership team to review and monitor the projects in play to implement the strategic plan.

Cynthia St. John: Support must be in place at the top. If the senior-most leader is not on board (i.e., CEO or president), then your work is likely to have traction only within the scope of the senior-most supporting leader (whether a division, function, etc.). Support and scope must be aligned for success. A true transformational journey is a long process that requires a lot of work and consistency... it is not for the squeamish or the fair-weather executive champion. Get the commitment and then ensure a strong and sustainable partnership between champion and lead change agent. When this is in place, it's magic; when it's not, it's menacing. Be committed to staying the course to avoid workforce trust issues or a "flavor of the month" culture.

JoAnn Sternke: I can only say what worked for me. The four areas that I believe you need to be very intentional and working on are people, building relationships, and listening to your key customers and stakeholders. People are first. If you don't start there, everything else will be misinterpreted by your stakeholders and customers. Plan. You must have a plan. I think a key part of our work is having a good planning process and then a good process to deploy that plan. We need to be results-driven. I think transformative organizations are measuring more than just student achievement, although that is the most important thing we measure and tend to. If we monitor things in other areas in our organization, we can funnel resources back to the classroom. Process is last because you adjust your processes to attain better results. You've got to know what results you are looking to improve and then adjust your processes accordingly. The advice that I have is stay the course and be very intentional about people, plan, results, and process.

Doug Waldorf: Create a plan. Ensure strong executive support and involvement.

WHAT ELSE WOULD YOU LIKE TO SHARE WITH OTHERS ABOUT YOUR JOURNEY TO EXCELLENCE?

I gave the interviewees an opportunity to share their insights beyond my questions. Some of those insights appear in previous chapters of this book. Listen and heed their advice. They have walked this path and seen the benefit to people, the organization, and themselves. Never, never, never give up. What if we took one step closer every day toward excellence?

RESPONSES

Genie Wilson Dillon: I have found that key areas of performance excellence are relevant to every aspect of life—not just in a work setting, but in personal life and relationships. Everything I have learned and experienced related to performance excellence has transformed my thinking about all aspects of life. I am a better person for this knowledge and experience using performance excellence thinking.

Vic Figurelli: Themes that would apply to nonprofits:

- Good leaders teach, rather than direct, and serve as role models for the behavior they're seeking.
- Leadership potential exists at all levels of the organization.
- Excellent organizations spend the effort to listen to their customers and to their people to learn how to serve them better.
- Excellence requires investment in the workforce to give them the skills, knowledge, and tools to carry out the organization's mission.
- Excellence requires that an organization builds and maintains a culture of trust, openness, and caring.
- Excellent organizations know where they are heading long term and use a strategic planning discipline to provide the pathway.
- Excellent organizations know their key business processes and work relentlessly to drive out waste and non-value added steps.
- Excellent organizations measure what's important to mission achievement.
- Excellent organizations perform!

Susan Fumo and Kelly Munson: Every organization needs a team like ours that can be in the center of the organization and support folks with the sole purpose of continuous improvement. I think that because organizations don't have that, they are trying to embed these processes amongst everything else they are doing in their job. This role brings the capacity to internally focus on continuous improvement always focusing on change, and having difficult conversations. Because we are so uniquely positioned, we don't report to anyone. We are positioned to build bridges between and across the district. Organizational health of the culture is very critical to change. Our sole purpose as a team is to transform our organizational health. A by-product of that work will be the success of our organization.

Frankie Jackson: When you have the capability to perform your craft well, the secret of excellence is in the details of the work which results in excellence. Excellence is the result of having the best of intentions with persistent effort, leadership, and skillful execution. Transformation is about doing the best in every moment so the next moment will be better than the last. The journey to excellence is at the heart of true visionary leadership with stretch goals, and most of all—HARD WORK. Performance excellence must be embedded in the organization as the standard where each staff member does their best to achieve it every day! My journey as a leader is to plan for it, model it, and motivate it throughout the organization.

Rick Rozelle: The journey is never ending. That's because the work is cyclical—you plan it, perform it, look at the results, adjust the plan, and begin again.

Cynthia St. John: It's well worth the time and toil when you step back to see the dramatic positive change over time. The impact touches everyone who touches the organization—whether employee, customer, supplier, or community. Success lifts all.

JoAnn Sternke: I don't think it (continuous improvement) needs to be complex. I think we know it is about people, plan, results, and process. Even though people don't always think about process intentionally, I think we know it's about people and mission first, but it's about how intentional we are in putting things in motion to tend to that or measure our success, our effectiveness in those areas. I think it's about being intentional and mindful about the work we do around p, p, r, p. I think it makes all the difference in an organization.

Doug Waldorf: Don't give up!

Note

1. S. Godin, *Tribes: We Need You to Lead Us* (New York: Penguin Books, 2008), 1.

Appendix

Action Plan Template

Goals or Objectives: *What is the strategic goal or objective linked to this action plan?*		
Action Plan Statement: *What are you going to do? Example: Develop a Voice of the Customer strategy*		
Project: *If this plan is part of an initiative state it here*	Link(s) to Strategic Plan:	
Leadership and Team		
Purpose: *Why do we need this plan? How does it help us achieve our goal?*		
Key Deliverables: *What will this plan produce? Products? Services?*		
Resources: *What internal resources do you need from other departments or units?*		

Measures	Performance Indicator(s)	Target(s):
In-Process	• *Leading indicators*	•
Outcome	• *Lagging indicators*	•

Timeline:	*What is the time frame for this action plan?*

Risks/ Success Factors	*What can threaten the accomplishment of this action? What will support the completion of the plan?*

Cost Overview	*How much will this cost?*	Potential Funding:	*What is the source of funding?*
Interdependencies:	*Who needs to contribute or be involved in this action plan from across the organization?*		

Action Plan Steps	Responsible	Progress	
		Start	Status
1.			
2.			
3.			

Agenda Template

Vision / Mission / Values
Date of the Meeting

Meeting Objective(s):

1.
2.

Logistics:		Members:	Meeting Guidelines
Date:			• Be respectful
			• Open sharing
			• One conversation at a time
Time:			• Mute phones
			• Take care of your own needs
Location:			• Stay on topic
			• Consensus
			• Silence means affirmation
Bring:			• ELMO
			(Enough Let's Move On)
Preparation Required:			• Have fun!

AGENDA ITEM	LEAD	START	DURATION
1)			
2)			
3)			
4)			
5)			

Meeting Evaluation
Plus/Delta

+	Δ (I)

Communication Plan

Successful change management requires a systematic approach to communicate, communicate, communicate. You can never over-communicate. However, what is communication? Effective communication requires four integrated components to achieve shared meaning:

- A clear message with enough detail to develop understanding
- A listener who trusts the sender of the message and will ask clarifying questions
- An appropriate delivery method that meets the needs of the sender and the receiver
- Content that connects with the listener

Authentic and intentional communication begins with a clearly written plan that includes these characteristics:

- Consistent, frequent, and through multiple channels
- Updated as quickly as the information is available
- Allows for questions, clarification, and input
- Clearly communicates the vision, the mission, and the objectives of the change management effort
- Two-way and real discussion—a conversation
- Opportunities for leaders or sponsors to spend time conversing one-on-one or in small groups with the people who are expected to make the changes
- Reasons for the changes so people understand the context, the purpose, and the need
- Pro-active
- Public review of the measurements that are in place to chart progress in the change management and change efforts

There are many ways to write a communication plan. The matrix that follows provides a simple approach to ensure that you have considered the key components for communicating the change effort.

Target Audience	Key Message	Delivery Date	Medium	Frequency	Messenger	Indicator(s) of Success

Measurement Plan

How will we know if the process is working?

A good measurement plan identifies key points in the process that will help us know how, or if, a process is working. Regular evaluation of process keeps the effort on track and helps to keep design choices in place.

To develop a plan to measure and evaluate the new process design:
1. Establish the criteria for measuring the success of the design
2. Identify what information is needed to measure the new design against the criteria
3. Determine how, when, where, and who will collect the information
4. Identify how the information will be used to monitor the new design during the transition period
5. Establish checkpoints for reviewing the progress using data
6. Act based on the information you obtain to maintain or improve the process.

Measurement Criteria	Information Needed	How to Collect	When to Collect	Who Collects	Actions
Completed ≤ 1 hr	*How long does it take to complete the process or step?*	*Measure the time from start to end*	*Sample three times per day*	*Roy Rogers*	*If the cycle time exceeds 1 hr three days in a row, determine cause*

Project Charter

Project Name:

Sponsor:	Team Lead:	Date:
Core Team:		
Extended Team:		

Vision, Mission, and Goals of the Organization *(How does this project support organizational strategy?)*
Problem Statement *(What is the problem? Are we meeting customer requirements? Is it an internal process issue?)*
Mission/Goal Statement *(State your goal or mission for the project in measureable terms using SMART.)*
Scope *(What is included in the project and what is being excluded from this project work? What are the project and process parameters, beginning, end?)*
Deliverables *(What products will the project work create?)*
Key Performance Measures (Metrics) *(What is your measurement system for analyzing the problem and controlling the solution?)*
Interdependencies/Support/Resources *(Who and what is needed to ensure success of the project?)*

Project Milestones/Action Items

Milestone	Responsible	Due	Status

Risk Assessment

A risk assessment identifies and evaluates potential events (possible risks and opportunities) that could affect the achievement of objectives, positively or negatively. Risks can occur in the external environment or within the organization. Determining the risk of any change is critical to the decision-making process and provides a proactive approach to prevention, detection, or response to risk.

If your project requires a more robust risk analysis, use a Failure Mode Effects Analysis (FMEA). This tool provides more specificity and a method to weight the potential effect of implementation. You can find an example on American Society for Quality website: http://asq.org/learn-about-quality/quality-tools.html.

What is the potential problem?	What would cause this problem?	What are preventative actions?	What are the contingent actions?

Roles and Responsibilities RACI Diagram

A roles and responsibilities chart clarifies relationships and establishes responsibilities for decisions, tasks, and activities to achieve desired project or process goals. RACI diagrams ensure that the right people are in place to handle complex processes and in the right role at the right time. RACI diagrams can be used to identify roles and responsibilities at multiple levels of the organization, for example across the organization, within a department, division, or within a work group. RACI diagrams also ensures there are no:

- Ambiguous roles and responsibilities among the process team
- Unclear work assignments
- Duplication of effort or unassigned tasks

RACI indicates if an individual is:

Responsible (R)—for and does the work Frequently, more than one R is needed for very complex or skill dependent process activities.

Accountable (A)—has policy and budgetary authority over the process. Only one A per task allowed. If the process fails, this individual will be held accountable.

Consulted (C)—reviews work and provides input as needed to ensure a successful outcome

Informed (I)—of the process step being completed. This is a one-way information flow, frequently used as a "get ready, here it comes" message. As the sender of the information, you do not expect a response.

Activity	Sponsor	Team	HR	IT	Finance	Others?

SIPOC Purposes and Instructions

Purposes:

- To provide an "at a glance" overview of a process
- To define the start and stop boundaries of a process (and project scope)
- To clarify relationships of the suppliers of inputs to the process
- To identify process customers (internal and external), and the process outputs that they seek
- To identify unintended wastes output by the process

The SIPOC analysis tool provides a big picture view of the important elements of the process. The team uses the SIPOC to gain a deeper understanding of the broader context or system in which the process occurs.

Begin by naming the process starting with a verb and define the beginning and end, scope, to remain focused on the specific process. To use the SIPOC tool, follow these simple steps:

- List the process Outputs and their Customers
  Identify the outputs of the process. What are you producing with the process? Is it a product, such as a report or service? Does this process feed into another process? Look at your outputs. Who is the customer of these outputs? Is it a person, a department, another process?

  When considering the outputs, keep in mind the needs and requirements of the customer. Do you know what they are? Do not guess requirements; obtain direct input from the primary users/recipients of the process. You will also want to identify internal customers to ensure that your handoff to their input meets requirements.

- List Suppliers and the Inputs they provide
  Identify what is needed for your process. These are the inputs of the process. Is it materials, people, a product? One way to think about the inputs is to ask, if I don't have this "whatever that is" we won't be able to complete or improve the process. Consider your requirements of the inputs. Have you communicated those requirements to the supplier of your input?

- Review the list of suppliers. Which suppliers are internal (within the organization or department) and which are external (outside the organization or department)? Another consideration is the team's ability to influence the supplier. If the input is not up to standard, can the team do anything to affect change in the input?

- Draw high-level flow chart of the process
  Identify major steps only. If you have more than 7 steps — you may be too detailed. You can add details to the full flowchart that you create later. Use simple verb-noun descriptions.

Follow-up

- Validate
  After you have completed the SIPOC, check with others who know the process well to ensure a clear understanding of the process and the inputs, outputs, and customers.

- Identify key measures
  When you have validated the completed SIPOC, identify the key performance indicators you will use to ensure:
  - Quality of the input and output
  - Capability of the process to deliver an output that meets customer requirements
  - Voice of the Customer feedback on whether the output met their requirement

Suppliers Who provides input for this process?	Inputs What do they provide?	Process Steps What are the 3-7 major steps of the process?	Outputs What is created by the process?	Customers Who will use the outputs of the process?

Stakeholder Analysis and Commitment Plan

The stakeholder analysis identifies the individuals or groups with an interest in a project that includes a rollout of change that will need the support of those who can make it succeed. A *stakeholder* is anyone with an interest or right in an issue, or anyone who can affect or be affected by an action or change. Stakeholders may be individuals or groups, internal or external to the organization. Stakeholders may also be primary customers.

Use the following questions to identify stakeholders, their interests, and their level of commitment.

Interests
- Who might receive benefits?
- Who might experience negative effects?
- Who might have to change behavior?
- Who has goals that align with these goals?
- Who has goals that conflict with these goals?
- Who has responsibility for action or decision?
- Who has resources or skills that are important to this issue/project?

Commitment
- Who must be committed to the implementation of the project?
- What is their level of support?
- What essential support to you need from them?
- Who will take what specific actions to gain their support?

The matrix below can help in capturing this information and developing actions to ensure that you have considered all stakeholders and their needs.
- List the stakeholders
- Identify their interest and/or level of support
- Rate their attitudes (A) toward the issue/project
  - a.   ++ very positive or very supportive
  - b.   + positive and/or supportive
  - c.   0 neutral
  - d.   – negative
- Rate their level of influence (I) as high (H), medium (M), low (L)
- Rate the level of support/commitment (S) needed from this stakeholder as high (H), medium (M), low (L)
- Specify who will take what actions to gain that support.

Your communication plan should include the key messages for each stakeholder group. Clear, frequent information relative to the change can assist in involving stakeholders and gain their support or commitment.

Stakeholder	Interest/Support	A	I	S	Who?

SWOT Analysis

A SWOT analysis guides you to identify the positives and negatives inside your organization (S-W) and outside of it, in the external environment (O-T). Developing a full awareness of your situation can help with both strategic planning and decision-making. The following questions serve as a guide for your analysis.

Strengths *(internal, positive factors)*	Weaknesses *(internal, negative factors)*

Strengths describe the positive attributes, tangible and intangible, internal to your organization. They are within your control.
 	
• What do you do well?
• What internal resources do you have? Think about the following:
   o *Positive attributes of people,* such as knowledge, background, education, credentials, network, reputation, or skills.
   o *Tangible assets of the organization,* such as capital, credit, existing customers or distribution channels, patents, or technology.
• What advantages do you have over your competition?
• Do you have strong research and development capabilities?
• What other positive aspects, internal to your business, add value or offer you a competitive advantage?
	Weaknesses are aspects of your business that detract from the value you offer or place you at a competitive disadvantage. You need to enhance these areas to compete with your best competitor.
	
	• What factors that are within your control detract from your ability to obtain or maintain a competitive edge?
	• What areas need improvement to accomplish your objectives or compete with your strongest competitor?
	• What does your business lack (for example, expertise or access to skills or technology)?
	• Does your business have limited resources?
	• What are you holding on to that is draining resources?

Opportunities *(external, positive factors)*	Threats *(external, negative factors)*
Opportunities are external attractive factors that represent reasons your business is likely to prosper. • What opportunities exist in your market or the environment from which you can benefit? • Is the perception of your business positive? • Has there been recent market growth or have there been other changes in the market that create an opportunity? • Is the opportunity ongoing, or is there just a window for it? In other words, how critical is your timing?	Threats include external factors beyond your control that could place your strategy, or the business itself, at risk. You have no control over these, but you may benefit by having contingency plans to address them if they should occur. • Who are your existing or potential competitors? • What factors beyond your control could place your business at risk? • Are there challenges created by an unfavorable trend or development that may lead to deteriorating revenues or profits? • What situations might threaten your marketing efforts? • Has there been a significant change in customer behavior? • What about shifts in the economy or government regulations that could reduce your revenue?

Tools_Facilitation Tools

Facilitation Tools

The list below represents just a few of the tools that the team leaders or facilitator may use to manage group discussions and decisions.

Brainstorming: Generate ideas in a short period of time. Record every idea without judgment.

Force-field analysis: Identifies opposing forces of change where the driving force must be stronger than the restraining forces.

Group Norms: Agree on the behaviors needed for effective team participation. Post them during the meeting or include them on the agenda.

Nominal group technique: Brainstorm ideas in a structured format that encourages participation of all members of the group.

Parking lot or issue bin: When issues come up that may not be relevant to the discussion but are important to a team member, have them post that question, concerns, or issue on a post-it note on a chart labeled "parking lot."

Plus/delta: Solicits feedback on the strengths (plus) and opportunities (delta) of a meeting or situation.

Additional resources:

> Peter R. Scholtes and Brian l. Joiner, *The Team Handbook,* 3rd ed. (Madison, WI: Oriel, 2003)
>
> Nancy Tague, *The Quality Tool Box* 2nd ed. (Milwaukee: ASQ Quality Press, 2005).

Seven Basic Management and Planning Tools

Seven Basic Management and Planning Tools

Affinity diagram: Organizes brainstorming ideas into groupings and common relationships.

Interrelationship diagram or digraph: Shows cause and effect relationships between factors in a complex situation.

Matrix diagram: Develops a grid using an x and y axis to prioritize or analyze multiple factors

Prioritization matrix: Prioritize factors using a matrix and weighting the elements based on importance.

Process Decision program chart: Breaks down tasks into a hierarchy using a tree diagram.

Tree diagram: A graphic organizer to break down broad categories into finer levels of detail.

Activity network diagram: Sequences a set of tasks and sub-tasks that may occur in parallel. It helps identify the critical path or longest sequence of events.

Seven Basic Quality Control Tools

Cause-and-effect diagram (also called Ishikawa or fishbone diagram): Identifies many possible causes for an effect or problem and sorts ideas into useful categories.

Check sheet: A structured, prepared form for collecting and analyzing data easily adaptable for a wide variety of purposes.

Control charts: Uses a basic run chart that includes statistically determined control limits.

Histogram: A graph showing frequency distributions, or how often each different value in a set of data occurs

Pareto chart: Shows on a bar graph which factors are more significant.

Run Chart: Plots a series of data over time.

Scatter diagram: Graphs pairs of numerical data, one variable on each axis, to look for a relationship.

Additional Resources:

Robert D. Boehringer, *The Process Management Memory Jogger™,* (Salem, NH: GOAL/QPC, 2008).

The Memory Jogger™: A Pocket Guide of tools for Continuous Improvement, (Salem, NH: GOAL/QPC, 2008).

American Society for Quality (ASQ), Learn about Quality, http://asq.org/learn-about-quality/quality-tools.html

Voice of the Customer

Customer Requirements and Metrics

The Voice of the Customer (VOC) table is a tool to capture and organize information about how the customer will use a product or service and what the customer views as critical to quality. The VOC assists in identifying key customer requirements and translates them into metrics. These metrics provide information and specific data on the extent to which the product or service meets customer needs and requirements.

Customer	Voice of the Customer	Key Customer Issue(s)	Critical Customer Requirement	Metric
Who is the customer?	*What does the Customer want from us?*	*What is the issue(s) that prevent us from satisfying our customers?*	*Summarize key issues and translate them into specific and measurable requirements*	*What are the measurable requirements?*

Index